Recalibrating the Quantit
Revolution in Geography

This book brings together international research on the quantitative revolution in geography. It offers perspectives from a wide range of contexts and national traditions that decenter the Anglo-centric discussions. The mid-20th-century quantitative revolution is frequently regarded as a decisive moment in the history of geography, transforming it into a modern and applied spatial science. This book highlights the different temporalities and spatialities of local geographies laying the ground for a global history of a specific mode of geographical thought. It contributes to the contemporary discussions around the geographies and mobilities of knowledge, notions of worlding, linguistic privilege, decolonizing and internationalizing of geographic knowledge.

This book will be of interest to researchers, postgraduates and advance students in geography and those interested in the spatial sciences.

Ferenc Gyuris is an associate professor of geography at ELTE Eötvös Loránd University, Faculty of Science, Institute of Geography and Earth Sciences, Department of Social and Economic Geography in Budapest, Hungary.

Boris Michel is a professor of geography at Martin Luther University in Halle, Germany.

Katharina Paulus is a PhD student at the Institute of Geography, Friedrich-Alexander University in Erlangen-Nürnberg, Germany.

Routledge Research in Historical Geography

This series offers a forum for original and innovative research, exploring a wide range of topics encompassed by the sub-discipline of historical geography and cognate fields in the humanities and social sciences. Titles within the series adopt a global geographical scope and historical studies of geographical issues that are grounded in detailed inquiries of primary source materials. The series also supports historiographical and theoretical overviews, and edited collections of essays on historical-geographical themes. This series is aimed at upper-level undergraduates, research students and academics.

Twentieth Century Land Settlement Schemes
Edited by Roy Jones and Alexandre M. A. Diniz

Resisting the Rule of Law in Nineteenth-Century Ceylon
Colonialism and the Negotiation of Bureaucratic Boundaries
James S. Duncan

Cold War Cities
Politics, Culture and Atomic Urbanism, 1945–1965
Edited by Richard Brook, Martin Dodge and Jonathan Hogg

Micro-geographies of the Western City, c.1750–1900
Edited by Alida Clemente, Dag Lindström and Jon Stobart

Earth, Cosmos and Culture
Geographies of Outer Space in Britain, 1900–2020
Oliver Tristan Dunnett

Recalibrating the Quantitative Revolution in Geography
Travels, Networks, Translations
Edited by Ferenc Gyuris, Boris Michel and Katharina Paulus

For more information about this series, please visit: www.routledge.com

Recalibrating the Quantitative Revolution in Geography

Travels, Networks, Translations

Edited by Ferenc Gyuris, Boris Michel and Katharina Paulus

LONDON AND NEW YORK

First published 2022
by Routledge
4 Park Square, Milton Park, Abingdon, Oxon OX14 4RN

and by Routledge
605 Third Avenue, New York, NY 10158

Routledge is an imprint of the Taylor & Francis Group, an informa business

British Library Cataloguing-in-Publication Data
A catalogue record for this book is available from the British Library

Library of Congress Cataloging-in-Publication Data
A catalog record for this book has been requested

ISBN: 978-0-367-64086-6 (hbk)
ISBN: 978-0-367-64087-3 (pbk)
ISBN: 978-1-003-12210-4 (ebk)

DOI: 10.4324/9781003122104

Typeset in Bembo
by Apex CoVantage, LLC

Contents

Contributors

Larissa Alves de Lira is a visiting professor at Universidade Federal de Minas Gerais (UFMG), Department of Geography in Belo Horizonte, Brazil.
Email: lara.lira@gmail.com

Trevor J. Barnes is a professor and distinguished university scholar of geography at the University of British Columbia, Department of Geography in Vancouver, Canada.
Email: tbarnes@geog.ubc.ca

Luke R. Bergmann is an associate professor of geography at the University of British Columbia, Department of Geography in Vancouver, Canada.
Email: luke.bergmann@ubc.ca

Ferenc Gyuris is an associate professor of geography at ELTE Eötvös Loránd University, Faculty of Science, Institute of Geography and Earth Sciences, Department of Social and Economic Geography in Budapest, Hungary.
Email: ferenc.gyuris@ttk.elte.hu

Matthew Hannah is a professor of geography at the University of Bayreuth, Faculty of Biology, Chemistry and Earth Sciences, Chair of Cultural Geography in Bayreuth, Germany.
Email: matthew.hannah@uni-bayreuth.de

Mariana Lamego is an associate professor in human geography at Universidade do Estado do Rio de Janeiro (UERJ) in Rio de Janeiro, Brazil.
Email: marilamego@gmail.com

Boris Michel is a professor of geography at Martin Luther University, Institute of Geosciences and Geography in Halle, Germany.
Email: boris.michel@geo.uni-halle.de

Olivier Orain is a researcher at Centre National de la Recherche Scientifique (CNRS)/French National Centre for Scientific Research in Paris, France.
Email: olorain@wanadoo.fr

Katharina Paulus is a PhD student at Friedrich-Alexander University, Institute of Geography in Erlangen-Nürnberg, Germany.
Email: katharina.paulus@fau.de

Matteo Proto is an associate professor of geography at the University of Bologna, Department of History and Cultures in Bologna, Italy.
Email: matteo.proto2@unibo.it

Guilherme Ribeiro is an associate professor of geography at Universidade Federal Rural do Rio de Janeiro (UFRRJ) in Rio de Janeiro, Brazil.
Email: geofilos@msn.com

Michiel van Meeteren is a lecturer in human geography at Loughborough University, Geography and Environment in Loughborough, UK, and an assistant professor in human geography at Utrecht University, Department of Human Geography and Spatial Planning in Utrecht, Netherlands.
Email: m.vanmeeteren@uu.nl

Matthew W. Wilson is an associate professor of geography at the University of Kentucky, Geography Graduate Program in Lexington, Kentucky, USA, and a visiting scholar at Harvard University, Center for Geographic Analysis in Cambridge, Massachusetts, USA.
Email: matthew.w.wilson@uky.edu

Gego

Reticulárea, 1969. Museo de Bellas Artes, Caracas.

Steel, iron, bronze, lead, aluminum, paint and fabric.

Variable dimensions.

Photo: Paolo Gasparini

Fundación Gego Archives.

1 Introduction

Recalibrating the quantitative revolution in geography

Ferenc Gyuris, Boris Michel, and Katharina Paulus

Line as human
means to express
the relation between
points, something
that is entirely abstract
in the sense of not
existing materially in nature.

Line as medium
indicates materially
the relation between
points in space,
expressing visually
human descriptive thought.

Line as object to play with.

Gertrud "Gego" Goldschmidt, 1960

Introduction

The image on the previous page of this book shows a photo of the 1969 installation *Reticulárea* (from the Latin *reticulum* or fine network) by the German-Venezuelan artist and Jewish refugee Gertrud "Gego" Goldschmidt. In a room, a group of people stand between hundreds of interconnected rods of stainless steel and anodized aluminum. Four people, only the faceless silhouettes of whom can be seen, gaze at the structure; they seem lost and disoriented. The metal rods form a web of complex networks and meshes. While there is a remarkable regularity in triangles of different sizes, there is also chaos and distortion. It is not clear whether all the elements form one interconnected structure or whether different structures coexist within this space. The web has neither a center nor any clear boundary, and it appears to be a structure composed of the surface without any content.

Being trained also as an architect, Gego engages in her installations with contemporary spatial thought. *Reticulárea* was one of many geometric artworks that she, who cultivated a dedicated interest in surfaces, structures, and geometry, described as "drawings without paper" (Gego cited in Held 2010, 9).

DOI: 10.4324/9781003122104-1

Many of these structures, such as hexagons, regular spatial patterns, and optimal networks, remind us of a way of seeing and modes of visualizing that began to dominate geography in the 1960s. Geometry, rational planning and order, a focus on structure, connections and lines became important for a new breed of geographers and a new mode of geographic thought in the post-WWII period. However, Gego's artistic work was critical, expressing "a growing skepticism toward ideal urban, geometric, and sculptural organizations" (Amor 2005, 125) that dominated the practice of urban planning in Venezuela during the 1960s and elsewhere. *Reticulárea* can thus "be understood as a conflictive linear geometric body, whose behavior was a dialogical response to Gego's architectural background and the local material conditions in which her work was produced," an artifact "upsetting the notion of architectural space as a container, and of line as the boundary of bodies" (Amor 2005, 125).

This work of art serves as a starting point to think about the multisided and multisited history of a geography that saw progress and its future in geometry and networks, in universal spatial laws, and in rational spatial planning. However, in the same way as *Reticulárea* "embraced a logic of displacement" (Amor 2005, 125), this history has more open borders and less clear centers than are acknowledged in most accounts. Thus, this history is a history of many connections of small and large networks of traveling ideas and people and the constant transformation and translation of these ideas and concepts.

Since the early 2000s, research on the history of science has become increasingly interested in the history of the mid-20th century. In particular, the rise of a new form of large-scale research, cold war rationality, and new disciplines and modes of thinking, such as cybernetics, drew an increasing interest of researchers from a wide range of fields (Erickson et al. 2013; Solovey and Cravens 2012; Oreskes and Krige 2014; Rid 2016). The mid-20th century has become "history" and a new and exciting field for academic economies of attention. Moreover, post-WWII science can also be considered a prehistory to present-day sociotechnological figurations of digitalization and datafication and of "what tech calls thinking" (Daub 2020). This history helps us understand our present moment. In addition, the massive optimism about planning and the future, which permeated scientific work in the post-1945 period, attracted considerable interest because it is atypical of or even alien to many mainstream contemporary approaches in social sciences, including geography.

However, the history of mid-20th-century science is told primarily not only as a history of cold war science but also as a history that almost exclusively focuses on the United States while often still implicitly claiming universality. While there is – at least from a European perspective – some justification for this recentering of the world's history, these narratives provide minimal or no space for Western Europe and the Soviet Bloc, as well as the Global South and the decolonized world, which are more than interesting case studies and peculiar variations in this story.

This narrative has also been true for the emerging research on the history of mid-century geography and the rise of spatial science, quantitative methodologies, and the use of digital computers for spatial analysis. The 1950s can justifiably be regarded as a time when the center of geography shifted from Europe to North America. In addition, related studies acknowledge the pioneering role of some non-US authors (Barnes and Abrahamsson 2017), such as Walter Christaller (1933; Ullman 1941) and Alfred Weber (1909), as well as early quantitative works from Torsten Hägerstrand (1957) and his school in Sweden (e.g., Taylor 1977). However, in spite of these works, a critical historiography for the global history of quantitative geographies, comparable to the historiography that has been elaborated for Anglophone geographical traditions during the last two decades, especially by Trevor Barnes (1998, 2001a, 2001b, 2004, 2008), is still missing.

This volume is aimed at contributing to a more international and decentered view and understanding of what is commonly referred to as the "quantitative revolution" in geography. Therefore, this book contributes to the ongoing discussions around the geographies and mobilities of knowledge (Jöns et al. 2017), notions of worlding (Müller 2021), linguistic privilege, decolonizing and internationalizing of geographic knowledge (Ferretti 2020, 2021; Schelhaas et al. 2020; Minca 2018; Husseini de Araújo 2018; Jöns 2018; Gyuris 2018). Within the history of geography, the term "quantitative revolution" has been utilized over the last couple of decades with regard to different shifts in diverse geographical, social, and academic settings, both by historiographers of geography and geographers who saw themselves as part of this "revolution." The first use of the term is often attributed to Ian Burton (Burton 1963), who in 1963 – one year after obtaining his PhD in geography from the University of Chicago and without any reference to a spatial or national context – stated that "in the past decade geography has undergone a radical transformation of spirit and purpose, best described as the 'quantitative revolution'" (Burton 1963, 151).

Given the wide range of spatial, historical, and theoretical contexts for which the term is applied, the concept is rather vague in analytical terms, and its widespread use blurs the remarkable diversity of on-the-spot realities and the complex and often contradictory relations between places and contexts. However, the act of framing these diverse phenomena as a "quantitative revolution" is an essential part of a strong and clear-cut narrative, which may be massively oversimplifying and even misleading, but has been proven to be very efficient over many decades in creating a firm identity for people who either claim themselves or are claimed by others to have been the key actors in introducing a "spatial science" approach to geography. This narrative has proven so capable of "surviving" in the long run that even alternative (e.g., critical and radical) approaches to geography, which emerged later along a conscious criticism of thinking about the discipline as spatial science, often overtook its highly monolithic and homogenizing understanding of "the quantitative revolution," just replacing the former glorious narrative with a critical interpretation. What was an emancipation from conservative regionalism and a shift toward a theoretical as well as applied geography from the perspective of many "quantifiers," thus became for the critics a technocratic endeavor without any real-world relevance that had to be overthrown by a revolutionary action (Harvey 2000). Most likely, nowhere was this more forcefully presented than in David Harvey's 1972 "Revolutionary and Counter Revolutionary Theory in Geography and the Problem of Ghetto Formation" (Harvey 1972).

Based on these observations, the aim of this volume is to challenge the simplistic narratives of the "quantitative revolution" to reveal the multifaceted realities that it hides and the diverse ways it influenced different academic contexts and individual careers. Our goal is to provide a much more pluralistic story with diverse actors in diverse places, with diverse personal and academic backgrounds and motivations, pursuing diverse goals, and thus, becoming involved in different ways in events that are subsequently regularly identified as "cornerstones" of a quantitative "turn."

Deconstructing clear-cut narratives of the "quantitative revolution"

Clear-cut narratives of the "quantitative revolution" are based on several claims or assumptions, which sometimes are made explicit, and they are regularly present implicitly:

- There was a group of "quantifying" scholars, who at a point in their life consciously and consistently became "quantitative," even if they did not necessarily start their career as such. The works that they produced from then onward were thus predominantly "quantitative." This finding is also true for people who later shifted to another approach, such as David Harvey (Harvey 1969) who shifted to Marxist geography (Harvey 1973) or Gunnar Olsson (Olsson 1968) who shifted to a more philosophical approach to geographic and cartographic reason (Olsson 1980). In certain terms, their "quantitative period" is still interpreted as thoroughly quantitative. In contrast to these people, their "Others" were "nonquantifiers," who did not make any significant contribution to "quantification" – either they did not conduct any research or they only authored some "minor" and "insignificant" works with quantitative tones. Thus, a person was either a "quantifier" or a "nonquantifier" at a specific point in time in his/her career. As revolutions only know revolutionaries and counterrevolutionaries, it hardly comes into question that someone may have simultaneously contributed to both quantitative and nonquantitative domains of the discipline.
- The "quantitative" in "quantitative revolution" is simultaneously an essential and discursive category. Scholars who were quantitative in terms of the works they produced also identified themselves as "quantifiers" and vice versa. The two sets match each other.
- "Quantifiers," or at least their most prominent representatives, constituted one academic "group." They were members of a common scholarly network with similar personal backgrounds (e.g., RAND in the USA), motivations, and goals. They were thus acting like a "vanguard," which is also reflected often by their nicknames, such as "space cadets" – a name the "quantifiers" at the University of Washington acquired derisively from the UCLA cultural geographer Joe Spencer (Barnes 2004, 572). This dualistic

self-identification was also linked to a generational divide in many cases (refer to Forest R. Pitts' interpretation of "the modernists" vs. "the dinosaurs," cited by Barnes 2004, 579).

- As a consequence of the previous points, the most important quantitative works were all the products of these "quantifiers."
- Geography and geographers before the "quantitative revolution" fell short of adequately addressing the key practical issues of the 1950s and 1960s, partly because complex new challenges emerged, which previous geographers had been unfamiliar with, and partly because "prequantitative" geography had a limited capacity even to reflect on controversial topics of its era. As a result, geography lost much of its previous prestige both in academia and societal discourse. For example, US geography *a lá* Richard Hartshorne, which could not reserve the image of a "competent" discipline during WWII decision-making, induced a sort of internal "fermentation," especially among a couple of young geographers (refer to Ackerman 1945; Schaefer 1953; Morrill 1984).
- The reason for such a decline in "prequantitative" geography was its claimed descriptive and "idiographic" nature. This discipline was not seeking general spatial regularities, did not create universally valid models, and lacked the kind of "sophisticated" mathematical-statistical methodology that could have enabled these models (Schaefer 1953). In other terms, it was "not modern" and not a science.
- The "quantitative revolution" was a revolution for the totally novel approach that it introduced, and more generally, the radical nature of the change that it initiated. Even if there had been a few "predecessors," such as Walter Christaller in the 1930s, they were "so much ahead of their time" that their views failed to be integrated into the mindset and academic discourses of their contemporaries in geography (refer to Bunge 1962). In addition, the "quantitative revolution" was considered not a gradual, moderate, and organic change in the history of geography that is repeated but a historically unprecedented shift to a new era. This interpretation was strongly embedded in the modernist and developmentalist thinking of the post-WWII decades, becoming tangible in the works of leading US- and UK-based economists such as Karl Polanyi (1944), Walter Lewis (1954), Simon Kuznets (1955), and Walt Rostow (1960) (Taylor 1993; Gyuris 2014). For all of them, the social changes they hypothesized or described were not a few among many in human history but "The Great Transformation," "the" turning point in the history of humankind, and "the" radical shift from "premodern" societies to "the" modern society.
- Geography's "quantitative revolution" started in Anglophone academia, especially in the USA, where it gradually spread throughout the world, with different "time lags" in different countries. This revolution was consistent with cold war developmentalist thinking, e.g., the stages of economic growth by Rostow, which implied that the whole world is proceeding along the same and single trajectory of "development," where the US is taking the lead in every sense and others have no better strategy than to "copy" the American model.

Remarkably, although this clear-cut narrative tells an easy-to-follow and politically powerful story of how "the quantitative revolution" occurred in geography, it falls short of providing a similarly clear-cut definition of what comprises a quantitative geography. Recollections from both "quantifiers" and "nonquantifiers" are rather common in referring to quantitative geographers' obsessions with "*theory*" and "*universal laws*," with "*numbers*" and "*data*," with *"sophisticated" statistical methods*, and with the pure elegance of *mathematical symbols*, *formulas*, and *figures*. They also sketch, at least implicitly but sometimes explicitly, the image of a full-fledged quantitative geographer as a *male* expert (refer to Hanson 1993; Barnes 2004) of *leading-edge computing devices*, who revolutionizes human–machine relations and constitutes a modern counterpoint to "classical" scholars. This approach includes a firmly applied focus, a personal devotion to solving "*practical problems*" instead of sitting in an "ivory tower" or just describing what could be "seen" in the field. "Real" quantitative geographers also seem to be painstakingly correct in the details but use this "virtue" to reveal "*universal laws*" and to present them via spatial *models* and *mathematized geodesign*. These geographers do not want to comment on peculiar cases; they want to show universal truths in their totality. They are pursuing these objectives by taking a *technocratic* attitude, claiming themselves objective, unbiased, and value-free – in prioritizing economic growth and technological development over all other stances that they find *per se* less important or even inferior. Furthermore, clear-cut narratives discuss quantitative geographers as conscious "revolutionaries" who are fully aware of their "historical calling" and praise their great predecessors, the "founding *fathers*" (instead of mothers) of their quantitative project.

Despite these uncertainties, the career of this clear-cut narrative did not end with the radical, humanistic, and feminist critique of quantitative geography emerging from the 1970s onward. These new approaches remarkably broke with the positive, or even outright glorious, narrative, for they started to criticize the features of quantitative geography of which the "quantifiers" were most proud. Nevertheless, many elements of the former homogenizing and totalizing narrative of the quantitative revolution survived in the radical critique, e.g., claimed firm boundaries between prequantitative geographies and quantitative geographies, and "quantifiers" versus "nonquantifiers." This survival may be explained by a complex set of reasons, two of which are worth considering at this point. First, these homogenizing and totalizing narratives of the quantitative geography, which once served as a strong cornerstone of the "quantifiers'" self-justification, now illuminated from a different perspective, could form a solid basis for criticism of the quantitative turn. Second, there are many signs that a considerable part of radical critics also had a relatively limited overview of the history of prequantitative geographies, and thus, could easily accept and keep many "nonnormative" or "denormativated" elements of the positivistic "mythology." In addition, the totalizing narratives of the "quantifiers" about quantitative geography as "the only scientific" geography and the discipline of the future, with all other approaches being irretrievably "obsolete," sometimes tended to be replaced with similarly totalizing narratives of the new and critical approaches implying that quantitative geography is "out of date." Therefore, there is a need to reveal the heterogeneous realities of the "quantitative revolution" to *recalibrate* its clear-cut narratives.

Chapters of this volume

This volume is a collection of studies from diverse historical and geographical backgrounds. In Chapter 2, Michiel van Meeteren explores the history of Dutch quantitative geography. Commonly described as a latecomer to the quantitative revolution, Dutch spatial science coincided with the societal transformation of the late 1960s. Although these changes profoundly altered Dutch human geography, lumping together this transformation with the quantitative revolution predominantly erased the memory of an earlier phase of Dutch quantitative geography in the 1950s. The chapter describes how quantitative methods took root in Dutch geography in these two waves in the 1950s and 1960s and discusses why the first wave has been substantially forgotten.

Mariana Lamego's study in Chapter 3 is concerned with a place-based narrative about the arrival of quantitative geography in Brazil in the 1960s and 1970s. Her chapter explores the materialized network of human bodies and nonhuman objects responsible for the production, circulation, and reception of quantitative geographies in Brazil. By focusing on two specific sites, the Brazilian Institute of Geography and Statistics (IBGE), which is located in Rio de Janeiro, and the more peripheral Rio Claro University, which is located in São Paulo state, the paper highlights the uneven geographies of knowledge and the quantitative geography in Brazil.

In Chapter 4, Guilherme Ribeiro examines the role of translations in the construction of quantitative geography in Brazil. His focus is on the two journals *Revista Brasileira de Geografia* and *Boletim Geográfico*; both are published by the IBGE. The paper discusses the relation between the publication of translated texts in these journals and the construction of a quantitative geography and points out that within the global circulation of knowledge, translation is a highly political issue.

With Matthew Wilson's contribution, Chapter 5 focuses on the United States, one of the most important places for experimentation with digital mapping. Following Howard Fisher at the Harvard Laboratory for Computer Graphics and Spatial Analysis, Wilson examines the sociotechnological arrangements that allowed new modes of cartographic experimentation with computers and the transformation of computers into machines of geo-visualization. He examines these sociotechnological interactions, not to tell a static intellectual history but to disrupt easy origin stories with all of the cul-de-sacs and failures, tenuous allies and adversaries, and the fragility of thought and action.

In Chapter 6, Ferenc Gyuris examines the quantitative geography in Hungary in a broader historical context from the interwar period in the 1920s and 1930s until the 2000s. He discusses the remarkable diversity of authors in Hungarian quantitative geography and regional science in terms of formal educational background, institutional affiliation, scholarly networks, and even age. In addition, he locates communist Hungary in the international academic context by highlighting the complex impact of "Western" and "Eastern" scholarly influences, including the mobilization of publications and scholars, on quantification in spatial research in Hungary.

Olivier Orain firmly situates the social and epistemological history of the French "géographie théorique et quantitative" within the political context of the events of May 1968. In Chapter 7, Orain analyzes the quantitative turn of French geography from its rise in the early 1970s to its "peak" in the mid-1990s. In his analysis of the "revolutionary" 1970s, it becomes clear that the quantitative actors remained a minority and no replacement of the Vidalian paradigm occurred. Only in the aftermath (1981–1996), conducted "spatial analysis" took the lead throughout French geography. Olivier Orain focuses on the social characteristics of the movement and the way French geographers created a rather singular quantitative geography by mixing their local academic tradition with what they appropriated from Anglophone discourses.

Chapter 8 retells the history of the quantitative geography in German geography according to a new way in which geographers were thinking about nature. While most accounts of the quantitative revolution in German geography focus on economic and social geography, Katharina Paulus and Boris Michel follow the emergence of the concepts of ecology and the ecosystem in German geography in their paper and show how these concepts transformed the holistic nature of regional geography into something that is quantified and calculable. In focusing on the beginnings of the quantitative theoretical paradigm in German geography, especially the work of Carl Troll, the paper explores how this new way of thinking paved the way for the later shift in social and economic geography.

In his paper, Boris Michel uses the city in Chapter 9 as a lens to explore the history of the quantitative revolution in German geography. He argues that it is worthwhile to read the history of the quantitative revolution considering not only new geographical theories but also the objects of geographical inquiry. The problem mid-20th-century geographers faced was that they no longer could disregard the city. However, taking "the city" serious would have undermined the dominant paradigm. The paper argues that for the German quantitative revolution, it was exactly the link between a new geographical thought and the city that brought both to the forefront of German geographical research.

In Chapter 10, Larissa Alves de Lira performs an in-depth analysis at the Brazilian Institute of Geography and Statistics (IBGE). Founded in 1938, IBGE became a central site for the modernization of statistics in Brazil. Lira's paper shows how numbers and counting advanced IBGE's production of visualization between 1938 and 1960 and evaluates the success of data accumulation and advancement of the mathematical point of view in geography just before the quantitative revolution.

In his contribution in Chapter 11, Matteo Proto presents the theoretical framework and the historical background that characterized the progress of quantitative geographical research in Italy during the 1950s. Employing case studies on research projects developed in the interwar period, he shows that applied geographical research was characterized early by its intermingling with nationalist goals. After WWII, many scholars who had previously cooperated with the regime considered Anglophone-inspired quantitative methods. They followed international trends in the development of the discipline but also sought a depoliticized approach to research as a new beginning in their politically contestable personal career.

The following two chapters cite individual geographers. In Chapter 12, Trevor Barnes and Luke Bergmann reexamine William Bunge's *Theoretical Geography*, arguably the most forceful and most radical publication of early Anglophone quantitative geography. However, while this publication has become one of "the classics" of the "canon" of quantitative geography, it had a difficult history. The book was published under a different title and funded privately and was not published in the US. In their chapter, Barnes and Bergmann describe the historical context of the publication as well as the three different versions of the text in which the publication was published.

In his Chapter 13, Matthew Hannah cites Peter R. Gould, one of the central, but less often discussed, advocates of the subsequent quantitative revolution. In the late 1970s, Gould began exploring forms of quantitative analysis that are compatible with a Heideggerian philosophy. Hannah interprets Gould's advocacy of Q-analysis as a form of mathematical representation, which avoids critiques by humanistic geographers of quantitative reductionism. The chapter also addresses current discussions of possible emancipatory uses of mathematical representation in human geography.

Instead of a conclusion by the editors and a futile attempt to establish one common denominator of these histories or to just repeat the diversity, the volume will be closed with a virtual discussion about the quantitative revolution's legacy in Chapter 14.

Acknowledgments

This book owes much to academic discussions at some related international conference sessions, where several coauthors of the current edited volume presented papers or contributed to a fruitful exchange of ideas. These events include the session "Histories of the Quantitative Revolution from a Different Perspective: Practical Implementation in Service of Political Agendas" at the 2018 Annual Meeting of the American Association of Geographers in New Orleans and several sessions at the annual Neue Kulturgeographie – New Cultural Geography conferences in Germany from 2017 to 2020. An indispensable part of the project was our international workshop on the histories of quantitative revolutions in geography in Kiel, Germany in September 2019. The latter event was enabled by the generous funding that we received from the Fritz Thyssen Foundation and the German Research Foundation (DFG). The research has also been supported by the János Bolyai Research Scholarship of the Hungarian Academy of Sciences. We thank Charlotte Liebel for her help in organizing the 2019 workshop and this publication. We also thank Fundación Gego for the generous permission to use Paolo Gasparini's photograph of Gego's "*Reticulárea*." We appreciate the assistance provided by Routledge and its copyeditors throughout the book project.

References

Ackerman, E. A. (1945): Geographic Training, Wartime Research, and Immediate Professional Objectives. In *Annals of the Association of American Geographers*, 35(4), 121–143.

Amor, M. (2005): Another Geometry: Gego's Reticulárea, 1969–1982. In *October*, 113(2), 101–125.

Barnes, T. (1998): A History of Regression: Actors, Networks, Machines, and Numbers. In *Environment and Planning A*, 30(2), 203–223.

Barnes, T. (2001a): "In the Beginning Was Economic Geography": A Science Studies Approach to Disciplinary History. In *Progress in Human Geography*, 25(4), 521–544.

Barnes, T. (2001b): Lives Lived and Lives Told: Biographies of Geography's Quantitative Revolution. In *Environment and Planning D*, 19(4), 409–429.

Barnes, T. (2004): Placing Ideas: Genius Loci, Heterotopia and Geography's Quantitative Revolution. In *Progress in Human Geography*, 28(5), 565–595.

Barnes, T. (2008): Geography's Underworld: The Military-Industrial Complex, Mathematical Modeling and the Quantitative Revolution. In *Geoforum*, 39(1), 3–16.

Barnes, T.; Abrahamsson, C. C. (2017): The Imprecise Wanderings of a Precise Idea: The Travels of Spatial Analysis. In Jöns, H; Meusburger, P.; Heffernan, M. (Eds.): *Mobilities of Knowledge*. Cham: Springer, 105–121.

Bunge, W. A. (1962): *Theoretical Geography*. Lund: Gleerup, 1st edition.

Burton, I. (1963): Quantitative Revolution and Theoretical Geography. In *Canadian Geographer*, 7(4), 151–162.

Christaller, W. (1933): Grundsätzliches zu einer Neugliederung des Deutschen Reiches und seiner Verwaltungsbezirke. In *Geographische Wochenschrift*, 1, 913–919.

Daub, A. (2020): *What Tech Calls Thinking: An Inquiry into the Intellectual Bedrock of Silicon Valley*. New York, NY: Farrar, Straus and Giroux.

Erickson, P.; Klein, J. L.; Daston, L.; Lemov, R.; Sturm, T.; Gordin, M. D. (2013): *How Reason Almost Lost Its Mind: The Strange Career of Cold War Rationality*. Chicago, IL: University of Chicago Press.

Ferretti, F. (2020): History and Philosophy of Geography I: Decolonising the Discipline, Diversifying Archives and Historicising Radicalism. In *Progress in Human Geography*, 44(6), 1161–1171.

Ferretti, F. (2021): History and Philosophy of Geography III: Global Histories of Geography, Statues That Must Fall and a Radical and Multilingual Turn. In *Progress in Human Geography*, OnlineFirst.

Gyuris, F. (2014): *The Political Discourse of Spatial Disparities: Geographical Inequalities between Science and Propaganda*. Cham: Springer.

Gyuris, F. (2018): Problem or Solution? Academic Internationalisation in Contemporary Human Geographies in East Central Europe. In *Geographische Zeitschrift*, 106(1), 38–49.

Hägerstrand, T. (1957): Migration and Area: Survey of a Sample of Swedish Migration Fields and Hypothetical Considerations on Their Genesis. In Hannerberg, D.; Hägerstrand, T.; Odeving, B. (Eds.): *Migration in Sweden: A Symposium*. Lund: C.W.K. Gleerup, 27–158.

Hanson, S. (1993): "Never Question the Assumptions" and Other Science from the Revolution. In *Urban Geography*, 14(6), 552–556.

Harvey, D. (1969): *Explanation in Geography*. London: Edward Arnold.

Harvey, D. (1972): Revolutionary and Counter Revolutionary Theory in Geography and the Problem of Ghetto Formation. In *Antipode*, 4(2), 1–13.

Harvey, D. (1973): *Social Justice and the City*. London: Edward Arnold.

Harvey, D. (2000): Reinventing Geography. In *New Left Review*, 4, 75–97.

Held, L. (2010): Linien, die die Wirklichkeit auflösen. Leben und Werk von Gertrude Goldschmidt, genannt Gego. In *Ila – Zeitschrift der Informationsstelle Lateinamerika*, 338, 8–10.

Husseini de Araújo, S. (2018): Theories from the South in Leading International Geography Journals? In *Geographische Zeitschrift*, 106(1), 50–60.

Jöns, H. (2018): The International Transfer of Human Geographical Knowledge in the Context of Shifting Academic Hegemonies. In *Geographische Zeitschrift*, 106(1), 27–37.

Jöns, H.; Meusburger, P.; Heffernan, M. (Eds.) (2017): *Mobilities of Knowledge*. Cham: Springer.

Kuznets, S. S. (1955): Economic Growth and Income Inequality. In *American Economic Review*, 45(1), 1–28.

Lewis, A. W. (1954): Economic Development with Unlimited Supplies of Labor. In *The Manchester School*, 22(2), 139–191.

Minca, C. (2018): The Cosmopolitan Geographer's Dilemma: Or, Will National Geographies Survive Neo-Liberalism? In *Geographische Zeitschrift*, 106(1), 4–15.

Morrill, R. L. (1984): Recollections of the "Quantitative Revolution's" Early Years: The University of Washington, 1955–65. In Billinge, M.; Gregory, D.; Martin, R. (Eds.): *Recollections of a Revolution: Geography as Spatial Science*. London: MacMillan, 57–72.

Müller, M. (2021): Worlding Geography: From Linguistic Privilege to Decolonial Anywheres. In *Progress in Human Geography*, 45(6), 1440–1466.

Olsson, G. (1968): *Distance, Human Interaction, and Stochastic Processes: Essays on Geographic Model Building*. Ann Arbor: University of Michigan.

Olsson, G. (1980): Birds in Egg, Eggs in Bird. In *Research in Planning and Design*. London: Pion, 7th edition.

Oreskes, N.; Krige, J. (Eds.) (2014): *Science and Technology in the Global Cold War*. Cambridge, MA: MIT Press.

Polanyi, K. (1944): *The Great Transformation*. New York, NY: Farrar & Rinehart.

Rid, T. (2016): *Rise of the Machines: A Cybernetic History*. New York: W.W. Norton & Company.

Rostow, W. W. (1960): *The Stages of Economic Growth: A Non-Communist Manifesto*. Cambridge: Cambridge University Press.

Schaefer, F. K. (1953): Exceptionalism in Geography: A Methodological Examination. In *Annals of the Association of American Geographers*, 43(3), 226–249.

Schelhaas, B.; Ferretti, F.; Novaes, A. R.; di Schmidt Friedberg, M. (Eds.) (2020): *Decolonising and Internationalising Geography: Essays in the History of Contested Science*. Cham: Springer.

Solovey, M.; Cravens, H. (Eds.) (2012): *Cold War Social Science: Knowledge Production, Liberal Democracy, and Human Nature*. New York, NY: Palgrave Macmillan.

Taylor, P. (1977): *Quantitative Methods in Geography: An Introduction to Spatial Analysis*. Boston: Houghton Mifflin.

Taylor, P. (1993): *Political Geography: World Economy, Nation-State and Locality*. London: Longman.

Ullman, E. (1941): A Theory of Location for Cities. In *American Journal of Sociology*, 46, 853–864.

Weber, A. (1909): *Über den Standort der Industrien*. Tübingen: J.C.B. Mohr.

2 In the footsteps of the quantitative revolution?

Performing spatial science in the Netherlands

Michiel van Meeteren

Introduction

The textbook account of the Dutch version of geography's quantitative revolution is clear-cut. The Netherlands 'lagged behind in the revolutionary mood' but caught up in the 1970s pretty rapidly (Van Hoof and De Pater 1982, 36; Knippenberg 2008, 61). This notion of 'standing in the footsteps of the American quantitative revolution' is informed by the biographical experience of a young generation of geographers arriving at the scene in the 1960s. The account of Dutch spatial science by one of these youngsters, Frans Dieleman (1942–2005), meticulously outlines the gradual adoption of quantitative and computational methods in the Dutch curriculum from the late 1960s onwards (Dieleman and Op 't Veld 1981). Dieleman and Op 't Veld's 'latecomer narrative' is corroborated, albeit viewed pessimistically, by a towering figure of the previous generation: Christiaan van Paassen (1917–1996). Van Paassen was present at the 1938 International Geographical Union (IGU) conference in Amsterdam where he witnessed Walter Christaller present his central place theory. According to Van Paassen, Dutch geographers were unimpressed (Borchert 1983; De Bruijne 1984; Van Paassen 1989).[1] He was also present as the sole Dutch on the Urban Geography Symposium in Lund, Sweden, in 1960 that was central to the international diffusion of the quantitative revolution (Van Meeteren forthcoming). Here he recalls that he could hardly comprehend what the mathematically oriented Americans were talking about (Van Paassen 1989). Eventually, Van Paassen (in De Bruijne 1984, 94) would lament the one-sidedness of the 'spatial paradigm' as it emerged in the 1970s. Thus, when both the Dutch quantifiers and its main detractor agree that the Netherlands was a latecomer to quantitative geography, we can consider the debate settled.

But then counter-narratives appear. Herman van der Wusten (2004, 49), who studied at the Municipal University Amsterdam (now University of Amsterdam) in the early 1960s, notes that Christaller had been part and parcel of the Amsterdam curriculum since at least the 1950s, and was considered 'okay but not particularly interesting'. He recalls his and his peers' surprise that in the US, Christaller was all of a sudden hailed as something new and exciting. And once one starts digging in pre-1960s Dutch geography, one finds studies

DOI: 10.4324/9781003122104-2

like Steigenga (1958) – who mentored Van Paassen at Utrecht university (Van Meeteren 2022) – that utilizes calculus to describe industrial decentralization tendencies in the Netherlands. This study is as sophisticated as the quantitative geography being contemporaneously written by William Garrison's tribe of quantitative revolutionaries at the University of Washington (Barnes 2004). Dieleman and Op 't Veld (1981, 147) do allude to older Dutch quantitative work, but nevertheless dismiss it. Even Christiaan van Paassen turns out to have stimulated quantitative work in the 1960s. His former student Bert van der Knaap, another quantifying pioneer in the Netherlands, credits Van Paaasen as an important mentor challenging him to increase his quantitative skills.[2] Moreover, Van Paassen's own work from that era can hardly be considered 'descriptive regional geography' and has a profound theoretical sophistication (Van Meeteren 2022).

That geography's historiography needs to understand the local context where history unfolds is by now accepted knowledge. Key episodes, such as the quantitative revolution or radical geography, articulate and mix with local traditions into profound variegation (Barnes and Sheppard 2019; Van Meeteren and Sidaway 2020). However, labels to demarcate specific conjunctures have effects and reverberate across the world. During the quantitative revolution in the US, several things came together: a rejection of an idiographic (individualizing) approach in favour of a nomothetic (generalizing, theory forming) one, an adoption of quantitative methods, the introduction of computers to assist calculation, an adherence to some form of the hypothetico-deductive method, a significant expansion of higher education, and generational change (Burton 1963; Gauthier and Taaffe 2002; Morrill 1987; Van Meeteren 2019a). The developments being described as 'revolution' signals that these changes took place during a short period. Outside the US, some of these developments occurred earlier, did not occur at all, or were spread out over a longer period. However, once a revolution is called in one place, it can become a rallying cry in others. People adopt terms like 'the quantitative revolution' to describe their own project. To borrow a phrase from economic geography, the idea of a 'quantitative revolution' as a stage through which geography evolves became performative (Barnes 2008). Dutch geographers, both revolutionaries and counter-revolutionaries, performed the quantitative revolution, mimicking the language and framing they learned from anglophone textbooks. The Dutch experience is made equivalent to the American episode, backgrounding potential differences. We can surmise that this is aggravated by the fact that the quantitative revolution came from the US, whose youth and consumer culture were strongly admired in the Netherlands of the 1950s and 1960s (Schuyt and Taverne 2004). As such, being dazzled by anglophone intellectual hegemony is partly self-inflicted (Van Meeteren 2019b). It might be the admiration of American achievement that make locals fawn at the American tradition while neglecting their own.

This chapter investigates the conundrum of inadvertently characterizing Dutch spatial science as 'lagging'. How profound was this earlier wave of quantitative geography that Dieleman and Op 't Veld mention, yet downplay? What things *did* change in the 1960s that made them experience a sense of revolution? The chapter draws on an extensive literature review, archival sources, and interviews and correspondence with involved Dutch geographers between September 2019 and December 2020. The argument illustrates how an uncritical application from historiographical concepts and demarcations, such as the quantitative revolution, can render local histories invisible. Key is that this rendering was done by Dutch geographers, who were so immersed in the American hegemonic presentation that they overlooked curating their own tradition.

Rudiments: the antebellum

Key to understanding Dutch geography is to comprehend its foundational parochial conflicts. In the Netherlands, an early decision was made in 1921 to split the human and physical geography curriculum (De Pater 2001). After the split, the Amsterdam Municipal University and Utrecht University geography departments descended in a five decades long rivalry of who was the true torchbearer of Dutch human geography (De Bruijne 1984). This same 1921 decision pushed economic geography towards a subject in economics degrees (Lambooy 1992). Willem Boerman (1888–1965, Figure 2.1), professor in economic geography, finds himself firmly established in Rotterdam's economics

Figure 2.1 Willem Everhard Boerman (1888–1965).

Source: Reproduced from Boerman (1930)

faculty. Boerman had been part of the losing faction opposing the human and physical geography split (Heslinga 1983) and co-authored a book on 'physical and mathematical geography' (Blink and Boerman 1919). Early on, Boerman (1926) developed innovative perspectives on relative space and a form of time-space convergence, referring to Alfred Weber and classical location theory. He heralded Christaller's dissertation (Boerman 1933) and was notably entrepreneurial, catapulting his students in power broker research positions in the Dutch state apparatus (Van Meeteren 2022). Boerman was well acquainted with Jan Tinbergen (1903–1994),[3] a junior colleague in Rotterdam who would become a key player in Dutch economic planning and the inaugural winner of the Nobel prize in economics. Boerman was editor in chief of the *Tijdschrift voor Economische Geographie*[4] during the 1930s, 1940s and 1950s. Back then, TE(S)G was a practical alternative to the official journal of the Dutch Geographical Society (De Pater 2009), and Boerman opened its pages to economic geography publications by practitioners, planners, economists, and engineers.

Pre-1960s innovations in Dutch human geography were driven by the discipline's engagement in spatial planning (Van Meeteren 2022). Even before geographers got involved in the 1930s, the emerging Dutch planning profession was consumed by a rift between architects, focusing on aesthetics, and engineers, interested in surveying before plan (De Ruiter 1980; Van der Valk 1983). The engineers considered accurate demographic projections a key technique to estimate housing demand. Quantitative models from demography were adapted to the regional level to facilitate surveys for expansion plans. Importantly, these practically minded engineers toned down the Malthusian overtones and positivism in demographic models, favouring a more pragmatic 'it gets the job done' epistemology (De Gans 1999). A key player here was Theo van Lohuizen (1890–1956) who was responsible for the survey work for the general expansion plan of Amsterdam (1935) (Van der Valk 1990). Van Lohuizen had proposed an export base model by the mid-1920s (De Smidt 1967) to calculate housing prognoses based on estimating propulsive employment. This model (Van Lohuizen and Delfgaauw 1935) was further developed together with economist (and trained economic geographer) Gerardus Delfgaauw (1905–1984) and gained widespread adoption (De Smidt 1967). However, the statistics necessary to make the calculations were unavailable to make use of the model in the 1935 general extension plan of Amsterdam,[5] underlining how a quantitative revolution is contingent on the availability of reliable data.

As there were not enough engineers willing to do the survey work in planning, this role was taken up by geographers in the 1930s (De Ruiter 1983; Van Meeteren 2022) who therefore became acquainted with this engineering and economist knowledge. The survey work produced by geographers in the 1930s was modelled on traditional regional geography (Stolzenburg 1984) and mathematics was limited to descriptive statistics (Knippenberg 2008), something commonly dismissed by engineers who lamented that geographers were unfocused and did too detailed unnecessary research in surveys (De Ruiter 1983). Geographers most exposed to quantitative methods usually had no university position, and it was often in the practice of survey work that scholars from the rivalling academic factions mingled, collaborated and befriended one another (Kruijt 1944). Meanwhile, these rivalries had become so intense that Amsterdam Municipal University professor Henri Nicolaas ter Veen (1883–1949) refused to participate in the 1938 IGU conference in Amsterdam that was organized by Boerman in cooperation with his Utrecht colleague Louis van Vuuren (1873–1951) (Heinemeyer in De Bruijne 1984, 109). Nevertheless, all rivals joined forces when they set up an inter-university research centre for applied research in 1941, the ISONEVO[6] (De Ruiter 1983). Boerman, Van Vuuren and Ter Veen actively catapulted their students to government agencies for statistics and planning, particularly in the jobs that count. Johannes Verstege (1912–1992) a student of Van Vuuren would become responsible for the Dutch census at the Dutch Central Bureau of Statistics (CBS) before becoming director of that agency in 1967 (Het Parool 1967). Future geography professors Hendrik Keuning (1904–1985) and Adriaan de Vooys (1907–1993) also worked at the CBS in the 1930s and 1940s (Knippenberg 2008). Things accelerate when the Nazi occupiers instil a top-down planning model in the Netherlands in 1941. The new *Rijksdienst voor het Nationale Plan* (RNP) and *Provinciale Planologische Diensten*[7] again form an employment pool for Dutch geography students (Van Meeteren 2022). Boerman's student George Zeegers (1911–1988) becomes director of research at the RNP (Boerman 1951), where the groundwork is laid for post-war reconstruction (Van Meeteren 2022). Perhaps the most iconic spatial legacy under German occupation is when Christaller's central place theory becomes instituted as a planning model in the newly reclaimed lands of the Noordoostpolder (Bosma 1993), a model which survives the war (Boyle et al. 2020, 94).

The first wave: quantitative applied geography between 1945 and 1960

Despite their controversial origins, the national planning institutions are largely retained after the end of the Second World War. The rapid post-war reconstruction coincides with the forestalled modernization and automobilization of the country (Van Meeteren 2022). In the final stages of the war, the geographers that worked at the CBS had participated in brainstorm sessions led by Jan Tinbergen on how applying mathematics could help in a government-led post-war reconstruction (Schuurmans and De Vries 1996, 20–21). This was reflective of an emerging technology-driven Fordist order in the post-war period where economic planning, mathematics and computers were venerated (Schuyt and Taverne 2004, chap. 4). Geographers, albeit proximate to key players such as Tinbergen, were ambivalent.

In 1946, George Zeegers gives a lecture where he pleads to reform the educational program to make geographers ready for a career in planning. He notes that geographers bring a lot to the table for survey research to bridge abstract theoretical and concrete contextual knowledge. However, he does argue that geographers ought not to be trained as 'walking encyclopaedias of geographical facts' and develop a pragmatic and practical attitude to applied research (Zeegers 1946). Around this time, two prominent central place studies are published. Keuning (1948), in one of his last publications as CBS employee before becoming the inaugural professor of Geography at the University of Groningen, publishes a Christaller-based central place categorization of the Netherlands. Also referring to Christaller is Amsterdam-trained Takes' (1948) study about the effect of the newly reclaimed lands (the Flevopolders) on regional central place systems. Johan Winsemius (1910–1964), another Amsterdam geographer who worked for the RNP in the 1940s and 1950s (Steigenga 1964), publishes another important quantitative study (Winsemius 1949). This study makes a complete survey of Dutch industrial geography applying Van Lohuizen's export-base method. Van Lohuizen himself is asked by the RNP to coordinate planning research, where he is to monitor that geography research does not become too expansive and remains relevant for planning applications (Van der Valk 1990). Van Lohuizen, who had already been teaching planning at the Amsterdam Municipal University, becomes a part-time professor at the Technical University in Delft in 1946. Geographers of the various clans are invited to mingle with the engineers in this educational environment (De Ruiter 1983), further hybridizing practices.

Mass planning in post-war reconstruction meant a high demand for speedy survey work. In 1949, a controversial survey of the city of Amersfoort is published (Klaassen et al. 1949). The lead author, Leo Klaassen (1920–1992)[8] is a student of Tinbergen and employs novel econometric methods to regional planning. Willem Steigenga (1913–1974), in his role as researcher for the city of Rotterdam and editor of TESG, writes a balanced review (Steigenga 1950), where he praises the study and admits that in the past, there was a lack of quantification, but he laments that the study excesses in '"veneration of numbers", . . . there is too much iconoclasm, too little reform' (Steigenga 1950). In TESG, a debate between Klaassen, geographers and engineers (Klaassen 1952; Van Aartsen 1953; Angenot 1953) ensues on the usefulness of quantitative methods to speed up the survey process. This debate is encouraged by TESG's editorial board (1952) that includes Boerman, Keuning, Steigenga and Zeegers. The other geographer in the debate, Van Aartsen (1953) agrees with Steigenga that although quantitative methods are useful, they should not lead to empiricism and number fetishism. Likewise, then census director Verstege, in his inaugural lecture (Verstege 1951) on 'social research and statistics' insists that mathematical research entirely modelled on the natural sciences will lead to an unjust society. Resultantly, social sciences need to alternate between generalizing and individualizing research. Similarly, in a Belgian lecture, Boerman (1950) lays out his philosophical foundations and argues that the central object of geographical research is understanding those relations, and processes, that generate geographical difference, bringing him close to a nomothetic position. Steigenga pushes for a disciplinary emancipation, where geographers do not only supply background numbers for urban designers but become 'social engineers' that develop theoretical models based on idealized theories of social change (Steigenga 1957). Together, these examples show that there was a theoretical and quantitative momentum in the 1950s, and geographers actively participated in the conversation. Their role, however, was to propose a nuanced interplay of nomothetic and idiographic approaches to the interdisciplinary dialogue. Positioning the Dutch applied geography vanguard in the foundational debate of American 1950s geography, they were more bullish on nomothetic geography than Hartshorne was but surely not as radical against idiographic geography as Schaefer (Barnes and Van Meeteren, in press).

Meanwhile, the Amsterdam geographers had come under the spell of American quantitative sociology that boomed during the early 1950s (Abbott and Sparrow 2007). Many cast away their sociographical[9] identity and become the founding generation of modern Dutch sociology (Van Doorn 1956). Consequently, quantitative sociological methodology, as presented in Lazarsfeld and Rosenberg's (1955) reader, is taught in the Municipal University Amsterdam geography curriculum from the mid-1950s. As regards institutionalization of quantitative methods in education, the Municipal University was more than a decade ahead of the other Dutch departments. This explains Herman van der Wusten's (2004) surprise about the revolutionary fuss coming out of the US, it had been their standard undergraduate curriculum way before Bill Bunge came barging in claiming unprecedented change in geographic thought.[10]

Nevertheless, more intense use of numbers slowly became overwhelming for practical research. The correspondence between Steigenga, Winsemius (geographers), Angenot and Van Lohuizen (engineers) on how to best calculate the concentration numbers for the export base studies for the RNP has been preserved. The handwritten proofs of theorem and counter-calculations span a hefty stack of paper.[11] The absence of computers in the Dutch 1950s social science context did seem to cap the quantitative momentum. In the 1958 lecture commemorating his retirement,[12] Boerman nevertheless argues that geographers and economists will need to work more closely together and that geographers have to master quantitative methods. He hopes his successor will have the proficiency to make that happen.

To conclude, in the first wave, it were particularly applied geographers who were engaging with quantitative methods, in a context where these were pushed by economists and engineers. Because geographers had become part of these networks in government research agencies in the 1930s and 1940s, they were in the position to plead for a nuanced geographical perspective on the use of these numbers. This mediating role of geography in qualified adoption of quantitative methods is distinctive from the US-based narrative (Barnes 2004; Burton 1963) of geography's quantitative revolution.

The second wave: Dutch geography in the 1960s

Whereas most action on the quantitative geography front in the 1950s was outside academia, the early 1960s finally see the applied geographers breaking in as new geography and planning institutes are staffed.[13] The ISONEVO is succeeded by SISWO, the Inter-university Institute for Social-Scientific

Research[14] in 1960 (Kouwe 1985). SISWO would become a key marginal space (Lorimer and Spedding 2002) for the development of quantitative geography in the Netherlands. SISWO's ranks are prominently filled with associates and students of the applied geography community (Kouwe 1985). One of the earliest research topics that SISWO sponsors is a large inter-university research programme on the future of inner cities. Project teams were formed spanning the boundaries between geography and economics and the different rivalling schools of geographers (Kouwe 1985, 6).

Now that geographers from different tribes that worked together in applied projects in the 1950s get university positions, the historical rivalries start to dissipate. Steigenga, a Utrecht geographer, becomes the inaugural professor in spatial planning in Amsterdam, unthinkable a few years earlier (Van der Valk 1983). Informal inter-university networks around SISWO and inter-university study groups on human geography emerge. In these circles, the latest theoretical developments in anglophone geography are discussed (e.g. Bours et al. 1964).

During the mid-1960s, Dutch geography is increasingly in dialogue with the anglophone literature in geography and regional science, and authors like Peter Haggett are actively debated in student circles.[15] Lambooy (1966) publishes a prize-winning article where he insists on coupling local theories with the latest English-language literature to rejuvenate regional geographical thinking. A modernized TESG publishes articles by young Anglophone quantifiers such as Leslie King (1962), Kevin Cox (1965), Wayne Davies (1965), Peter Gould (Gould and Leinbach 1966) and Ron Johnston (1966). TESG was a safe haven for spatial science at a time when the established US geography journals were reluctant to publish quantitative and theoretical geography (Barnes 2004).[16]

Inner city research, which had been independently continuing at Amsterdam Municipal University[17] intensified engagement with urban geographical theory (Heinemeyer et al. 1967, Preface). The 1964 IGU conference in London sets off a chain of events tying-in Dutch geography internationally. Based on contacts developed with Torsten Hägerstrand and others,[18] the Amsterdam Municipal University organizes in 1966 an international conference on 'urban core and inner city' (Heinemeyer et al. 1967) featuring Hägerstrand, Peter Hall, Allan Pred and Gunnar Olsson. Herman van der Wusten, still a student-assistant at the time, recalls that Olsson subscribed them to the mailing list of the underground MICMOG mimeographed working papers[19] (Barnes 2004). In the slipstream of these contacts, Van Paassen spends time as a visiting professor in Lund,[20] Sweden, the start of a lifelong friendship with Torsten Hägerstrand (Van Meeteren 2019c).

While the curriculum and networks internationalize, geography enters a period of rapid expansion. During the 1950s and early 1960s, geography and planning departments were still cosy small-scale affairs and growth in geography students had been cushioned by the establishment of new departments (De Pater 1999, 21). Although institutes were somewhat hierarchical, they were also organized informally: there was no study guide, people would just wander in and out of lectures.[21] From the mid-1960s onwards, departments started to grow and professionalize, a development that would accelerate in the 1970s (De Pater 1999, 29). When the student numbers increased, there was scope to hire new staff and deepen the division of labour among staff. SISWO was a particularly popular venue to acquire staff from. Gerard Hoekveld (1934–2011) was hired from SISWO by the Amsterdam Free University in 1967 (De Pater 1998). SISWO's director, Piet Kouwe (1928–1997), would eventually become professor in quantitative methods in Nijmegen in 1969 (Kouwe 1988).

When departments expanded, quantitative skills became a hiring criterium. Marcus Heslinga (1922–2009) (Kouwenhoven 1984, 60), the leading professor of human geography at Amsterdam Free University, had first-hand witnessed the generational culture war that quantitative geography engendered in the UK. He encouraged the Free University's new hire, Gerard Hoekveld, to embrace quantitative geography intending to prevent intergenerational rifts. In Utrecht, there was a similar 'peaceful' adoption where newly hired staff members were encouraged to travel and learn quantitative methods from abroad (Van Ginkel et al. 2020, 102–103).

Junior staff members (Jan Lambooy at the Free University) or even student-assistants (Joost Hauer and Bert van der Knaap in Utrecht) were assigned statistics teaching because they had affinity with quantitative methods. The Utrecht (former) student-teachers are able to secure access to the university's Elektrologica EL X 8 computer where the first computational experiments in Dutch geography are conducted in the late 1960s.[22] The quantitative momentum was further stimulated through visiting scholarships to the US, particularly at Amsterdam Free University. Gerard Hoekveld travels to the US in 1967 and becomes convinced of the quantitative momentum (De Pater 1998). He then stimulates his assistants to apply for scholarships.[23] His student Frans Dieleman follows in the same year, starting a lifelong friendship with William Clark at Wisconsin-Madison (Clark 2005). Two years later, Jan van Weesep travels to Wisconsin-Madison in Dieleman's footsteps. In the US, the young Dutch geographers do not only learn the importance of utilizing computers but also get socialized in the American narrative about the quantitative revolution as a fundamental break in geographical scholarship. When the Free University scholars return, it is with a firm conviction about the American quantitative revolution, and the key role learning computer programming plays in this.[24] In the same period, Utrecht University's Joost Hauer conducts visits to study quantitative geography curricula in Lund (with Torsten Hägerstrand) and Göteborg (Olof Wärneryd) in 1968 and to Bristol (with Peter Haggett and David Harvey) in 1969.[25]

Meanwhile, Dutch geography and planning departments actively started to recruit for mathematical chops outside the geographical discipline. In Amsterdam, Anneke Hakkenberg, a mathematician who previously worked at a physics institute is hired by Steigenga in 1967 to teach quantitative methods (Figure 2.2). Steigenga handed her a copy of Peter Haggett's *Locational Analysis in Geography* (1965) with the expectation to integrate linear programming in the spatial planning curriculum.[26] At the Amsterdam Free University, convinced by what

Figure 2.2 A newly hired Anneke Hakkenberg (3rd right, seated) discusses spatial models at Willem Steigenga's (left) young planning institute at Amsterdam Municipal University. A (staged) 1967 or 1968 promotional photo showcasing the institute's ambition to immerse in applied quantitative geography.

Source: Van der Valk (1983, 118); reproduced with permission by Arnold van der Valk

he saw in the US, Gerard Hoekveld also pushes to hire a mathematician, Rinus Deurloo, in 1968 to further professionalize quantitative methods teaching.[27]

In the late 1960s, SISWO decides to bring together all the methods teachers in the Dutch academic social sciences to exchange teaching notes and ideas. This meeting helps ignite a spark that ultimately leads to the formation of an 'inter-university working group of quantitative methods' in 1968 (Kouwe 1985). In this working group, the young Dutch quantitative geographers would frequently come together, organized by Ad Goethals (1940–2007), a former student of Anneke Hakkenberg who after graduation was hired by SISWO. It is the experience of this SISWO group that organizes a study day in 1971 (Dieleman et al. 1971), that largely informs Dieleman and Op 't Veld's (1981) account of Dutch spatial science.

The SISWO working group starts out as a bimonthly reading group, where they would read the latest handbooks on quantitative geography which were rapidly being published in the UK and US. After discussing Peter Haggett's work as a baseline,[28] they would read Leslie King's *Statistical Analysis Geography*

(1969), David Harvey's *Explanation in Geography* (1969) and Adams, Abler and Gould's *Spatial Organization* (1971).[29] Inspired by these books and his travels, Hauer's (1971) summary of quantitative geography largely follows this Anglo-American reading of the 'quantitative revolution'.

Apart from working through handbooks, they helped one another mastering methods and computer programming.[30] The group also contributes to a research project on 'economic health' of regions (see Van der Knaap 1971), utilizing the latest in computational techniques. Rinus Deurloo recalls that it was somewhat of a 'proof of concept' exercise to show the wider geographical community what analytical worlds would become possible through these computational methods.[31]

The early years of the SISWO working group, which would survive for decades, coincide with seismic shifts at Dutch universities. As the result of student uprisings, particularly the occupancy of the executive building of the Amsterdam Municipal University in 1969, the university governance system is radically reformed (Schuyt and Taverne 2004, 299–304). The power of the professor as sole decider of the curriculum gets replaced by a democratic system where the student body obtains a significant voice. This democratization of universities is a generational watershed for university staff. The young quantifiers are better able to identify and cope with new student demands. And, as Hauer (1994, 698) recalls, quantitative geography felt for them as a way to 'break through' the hierarchies as the senior professors did not 'get' the language of formal mathematical models. Meanwhile, it is the older generation, such as Steigenga and Van Paassen, who face difficulty adapting to the new situation. Having themselves grown up in small-scale mentoring relationships between professor and prodigy, they now all of a sudden face masses of students determining what they have to teach. Whether fair or not, these professors were perceived by students as exactly the kind of institutions that they were revolting against.[32] Then, from the early 1970s onwards, the Anglophone quantitative new geography steadily diffused, including computational methods, along Dutch universities as meticulously set out by Dieleman and Op 't Veld (1981).

Discussion

The aforementioned narrative essentially tells a story of continuity and change. There developed a first wave of innovative Dutch quantitative geography with theoretical sophistication in the 1950s, in the realm of applied post-war reconstruction research. This wave was ultimately limited by the unavailability of modern computation and insufficient uptake at universities. Because it was applied, many publications never escaped the world of government reports and only incidentally made an international splash. Dutch academic geography in the 1950s, meanwhile, has been described as 'splendid isolation' where institutes in conflict primarily tended to their own projects and were reluctant to publish (Van Ginkel 1994; Van Ginkel et al. 2020). Moreover, some did not think highly of what happened in the applied world, and there were serious conflicts whether this planning work was really worthy of academic geography (De Pater 1999, 110). This scepticism by formal academia helps explain why the earlier 1950s innovations are so poorly documented. The most prominent quantitative pioneer, Willem Boerman, was a professor at an economics faculty that had argued on 'the wrong side' of the human-physical geography split that is regarded emblematic of Dutch geography (De Pater 2001). That made it difficult to fit him into Dutch geography's self-narrative. Moreover, much progress was made in marginal spaces outside the ivory tower such as SISWO whose achievements got lost in the interdisciplinary archival void. When a rejuvenated push for quantification arrives in the mid-1960s, it largely follows – performs – the scenario set out in the anglophone textbooks.

Rhetoric aside, it is nevertheless important to note that despite obvious American influence, some of the first wave sensibilities, including its healthy scepticism on the limits of quantification, did reproduce themselves. Both Joost Hauer (1967) and Anneke Hakkenberg (1969) published scathing reviews of Peter Haggett's (1965) *locational Analysis in Geography*. Hakkenberg (1969), as trained mathematician, points out that the kind of theoretical inferences made by Haggett cannot be backed up by his amateurish mathematics. In her article, she dismisses the naive empiricism of Zipf and Social Physics and pours cold water over many of the ontological claims of quantitative geography. Although Hakkenberg acknowledges her position was more purist than that of the other SISWO working group participants,[33] Van der Knaap also recalls that for him, mathematics always remained a means to an end. It had to be subservient to theoretical ideas that underpinned the operationalization of research problems.[34] It is telling that the Dutch quantifiers eventually adopt a Belgian term for the quantitative revolution: 'the new orientation in geography' (e.g. Van Hoof and De Pater 1982), based on their (SISWO-induced) contacts with Ghent geographer Pieter Saey. Saey (1968) formulated his new orientation in a clear admiration of American spatial science but like Lambooy (1966), he did so in a way that was consistent and compatible with the relational and theoretical schemas developed by people like Boerman and Van Paassen. When Piet Kouwe, hired from the SISWO in 1969 to Nijmegen to introduce quantitative methods, held his retirement speech in 1988, he called the quantitative momentum of the 1960s and 1970s a 'methodological intermezzo'. Spatial science had been an intermezzo that modernized Dutch human geography, but ultimately reached its limits and was subsumed in a continuing pragmatic, pluriform and applied geographical tradition (Kouwe 1988).

To conclude, the Dutch first wave 1950s quantitative turn was not a revolution. Compared to the US, it was more gradual and incomplete. The US's 1950s expansion of the university system that had fuelled the original quantitative revolution (Van Meeteren 2019a) only started happening in the Netherlands in the 1960s. Likewise, the Netherlands was late in introducing computers to geographical research. The social revolutionary developments of 1968 and 1969 radically reformed the social relations at university and would contribute to the fading away of the pedagogy of 'catalogues of geographical facts' (Van Westrhenen and Dijkink 1982). Thus, there are plenty of revolutionary changes that happened in the Dutch geography in the 1960s that make it feel like the country was standing in the footsteps of the US. Yet both 'quantitative' and 'theoretical' approaches to geography, considered the core of the American revolutionary moment (Morrill 1987), had a significant older foothold in Dutch geography. The theoretical implications of mathematical practice in geography and the tension between idiographic and nomothetic research had been digested in the first wave already. Moreover, sensibilities learned in the first wave did trickle down in the second, making that when Dutch quantitative geography was finally codified in a textbook, it hardly had a strong positivist theoretical signature (Hauer and Van der Knaap 1973). For better or for worse, it is historically false to say that Dutch quantitative geography stands in the footsteps of the American quantitative revolution. Nevertheless, it is undoubtedly the case that the enormously influential Anglo-American quantitative stepped on Dutch geography and made a daunting impression.

Acknowledgements

In addition to the interviewees, I would like to thank Trevor Barnes, Ben de Pater, Joost Hauer, Arnold van der Valk, Michel van Hulten and the book editors for valuable feedback and suggestions.

Notes

1 Van Paassen's assessment will have been coloured by him being a student of the Utrecht School. Rotterdam's Willem Boerman, the main organizer of the 1938 IGU conference, was an early central place theory enthusiast (Boerman 1933).
2 Interview Bert van der Knaap 25–01–2020.
3 The correspondence between Tinbergen and Boerman preserved in the online archive 'The Tinbergen Letters' https://tinbergenletters.eur.nl/theletters/ (last visited 5 October 2020) suggests both a close personal and professional relation.
4 From 1948 onwards *Tijdschrift voor Economische en Sociale Geografie (TESG).*
5 Interview GTJ Delfgaauw by Peter de Ruiter, 1981, (partial) transcript available at Archive: Het Nieuwe Instituut, Lohuizen, Th. K. (Theodoor Karel) van/Archief LOHUd25.
6 The acronym ISONEVO stood for 'Instituut voor Sociaal Onderzoek van het Nederlandse Volk' (translated as 'Institute for Social Research of the Dutch People').
7 The RNP's wartime record is a source of controversy which cannot be elaborated here for space constraints. Van Meeteren (2022) refers to the relevant literature.
8 Klaassen, together with Jean Paelinck would later become one of the leading figures of the Dutch chapter of the Regional Science Association (Lambooy 1992).
9 Amsterdam human geographers called themselves 'sociographers' in this era (see De Bruijne 1984).
10 Interview Herman van der Wusten 12–09–2019.
11 Archive: Het Nieuwe Instituut, Lohuizen, Th. K. (Theodoor Karel) van/Archief LOHUd231
12 Archive: M. W. Heslinga, Collectie HDC Protestantse Erfoed nr 194, Vrije Universiteit Amsterdam, folder 165.

13 The University of Nijmegen starts a geography programme in 1958, The Free University Amsterdam in 1961, (De Pater 2001); The Municipal University in Amsterdam starts the first department in urban and regional planning (Planologie) in 1962 (Van Meeteren 2022).
14 Interuniversitair Instituut voor Sociaal-Wetenschappelijk Onderzoek.
15 Herman van der Wusten, personal communication 16–12–2020.
16 Kevin Cox, personal communication 17–02–2018.
17 Michel van Hulten, personal communication 10–09–2020.
18 Michel van Hulten, personal communication 10–09–2020.
19 Interview Herman van der Wusten 12–09–2019, personal communication 15–12–2020.
20 Letter Van Paassen to Hägerstrand 18–02–1966. Torsten Hägerstrand papers, Lund University, Box 42.
21 Interviews Herman van der Wusten 12–09–2019, Jan Lambooy, 13–09–2019, Jan van Weesep, 29–09–2019.
22 Joost Hauer, personal communication 15–09–2020; Interview Bert van der Knaap 25–01–2020.
23 Interview Jan Van Weesep 29–09–2019.
24 Interview Jan Van Weesep 29–09–2019; Interview Rinus Deurloo 28–01–2020.
25 Joost Hauer, personal communication 15–09–2020.
26 Interview Anneke Hakkenberg 26–01–2020.
27 Interview Rinus Deurloo 28–01–2020.
28 Interview Anneke Hakkenberg 26–01–2020.
29 Interview Bert van der Knaap, 25–01–2020; Joost Hauer, personal communication 15–09–2020.
30 Interview Rinus Deurloo 28–01–2020.
31 Interview Rinus Deurloo 28–01–2020.
32 Interviews Herman van der Wusten 12–09–2019, Jan van Weesep, 29–09–2019; Arnold van der Valk 09–10–2020; see also Van der Valk (1983).
33 Interview Anneke Hakkenberg 26–01–2020.
34 Interview Bert van der Knaap 25–01–2020.

References

Abbott, A.; Sparrow, J. T. (2007): Hot War, Cold War: The Structures of Sociological Action 1940–1955. In Calhoun, C. (Ed.): *Sociology in America: A History*. Chicago: University of Chicago Press, 281–313.

Abler, R.; Adams, J. S.; Gould, P. (1971): *Spatial Organization: The Geographer's View of the World*. Eaglewood Cliffs: Prentice Hall.

Angenot, L. H. J. (1953): De Mathematische Methode in het Sociaaleconomisch en Sociaal Onderzoek. In *Tijdschrift voor Economische en Sociale Geografie*, 44, 92–93.

Barnes, T. J. (2004): Placing Ideas: Genius Loci, Heterotopia and Geography's Quantitative Revolution. In *Progress in Human Geography*, 28(5), 565–595.

Barnes, T. J. (2008): Making Space for the Economy: Live Performances, Dead Objects, and Economic Geography. In *Geography Compass*, 2(5), 1432–1448.

Barnes, T. J.; Sheppard, E. (2019): *Spatial Histories of Radical Geography: North America and beyond*. Chichester: Wiley.

Barnes, T. J.; Van Meeteren, M. (in press): The Great Debate in Mid-Twentieth-Century American Geography: Fred K. Schaefer vs. Richard Hartshorne. In S. Coen; Rosenburg, M.; Lovell, S. (Eds): *Routledge Handbook of Methodologies in Human Geography*. London: Routledge.

Blink, H.; Boerman, W. E. (1919): *Wis-en Natuurkundige Aardrijkskunde*. Groningen: Wolters-Noordhoff.

Boerman, W. E. (1926): Wegen. In *Tijdschrift voor Economische Geographie*, 17, 405–417.

Boerman, W. E. (1930): Bedrijf en Economische Geografie. In *De Industrieele Gids*, 14(5), 66–68.

Boerman, W. E. (1933): Review: W Christaller. Die Zentrale Orte in Süddeutschland. In *Tijdschrift voor Economische Geographie*, 33, 355–356.

Boerman, W. E. (1950): Economische Geografie en Haar Eigen Centrale Probleem. In *Bulletin De La Société Belge d'Etudes Géographiques/Tijdschrift Van De Belgische Vereniging Voor Aardrijkskundige Studies*, 19(2), 86–98.

Boerman, W. E. (1951): Het Nut van Economisch-geografisch Structuuronderzoek. In *Bulletin De La Société Belge d'Etudes Géographiques/Tijdschrift Van De Belgische Vereniging Voor Aardrijkskundige Studies*, 20(1), 22–36.

Borchert, J. G. (1983): Geography Across the Borders: On the Relationship between Urban Geography in the Netherlands and Germany. In *Tijdschrift voor Economische en Sociale Geografie*, 74(5), 335–343.

Bosma, K. (1993): *Ruimte voor een Nieuwe Tijd*. Rotterdam: NAI.

Bours, A.; Lambooy, J. G.; Van Hulten, M. (1964): *Mededelingen van de Studiegroep voor Wetenschappelijke Sociale Geografie, nr. 1*. Mimeo, copy available at Utrecht University Library.

Boyle, M.; Hall, T.; Lin, S.; Sidaway, J. D.; Van Meeteren, M. (2020): Public Policy and Geography. In Kobayashi, A. (Ed.): *International Encyclopedia of Human Geography*. Amsterdam: Elsevier, 2nd edition, volume 11, 93–101.

Burton, I. (1963): The Quantitative Revolution and Theoretical Geography. In *The Canadian Geographer/Le Géographe Canadien*, 7(4), 151–162.

Clark, W. A. V. (2005): Obituary: Frans Dieleman. In *Environment and Planning A.*, 37, 951–952.

Cox, K. R. (1965): The Application of Linear-Programming to Geographic Problems. In *Tijdschrift voor Economische en Sociale Geografie*, 56(6), 228–236.

Davies, W. K. D. (1965): Some Considerations of Scale in Central Place Analysis. In *Tijdschrift voor Economische en Sociale Geografie*, 56(11), 220–227.

De Bruijne, G. A. (1984): Gesprekken Met Prof. dr Chr. van Paassen en Prof. dr. W.F. Heinemeijer. In Kouwenhoven, A. O.; De Bruijne, G. A.; Hoekveld, G. A. (Eds.): *Geplaatst in de Tijd. Liber Amicorum M. W. Heslinga*. Amsterdam: Vrije Universiteit, 77–128.

De Gans, H. A. (1999): *Population Forecasting 1895–1945: The Transition to Modernity*. Boston: Kluwer Academic Publishing.

De Pater, B. (1998): Geograaf uit Roeping. In Hauer, J.; De Pater, B.; Paul, L.; Terlouw, K. (Eds.): *Steden en Streken: Geografische Opstellen voor Gerard Hoekveld*. Assen: Van Gorcum, 8–21.

De Pater, B. (1999): *Een Tempel der Kaarten*. Utrecht: Faculteit Ruimtelijke Wetenschappen Utrecht University.

De Pater, B. (2001): Geography and Geographers in the Netherlands Since the 1870s: Serving Colonialism, Education, and the Welfare State. In Dunbar, G. S. (Ed.): *Geography: Discipline, Profession and Subject since 1870*. Dordrecht: Kluwer, 153–190.

De Pater, B. (2009): From a Magazine for "Practical Gentlemen" to an Academic Journal: One Hundred Years of TESG. In *Tijdschrift voor Economische en Sociale Geografie*, 100(1), 5–19.

De Ruiter, P. (1980): Over de Opkomst van de Moderne Stedebouw en Planologie (1870–1945). In *AKT*, 4(2), 14–27 and 4(3), 11–21.

De Ruiter, P. (1983): *Stedebouw Onderwijs 1900–1945. Over de Voorgeschiedenis van het Onderwijs in Stedebouwkunde, Landschapsarchitectuur en Planologie*. Den Haag: NIROV.

De Smidt, M. (1967): Stuwend en Verzorgend: Een Verkenning van de Ontwikkeling der Konceptie. In *Bulletin van het Geografisch Instituut Universiteit Utrecht*, 4, 7–40.

Dieleman, F. M.; Hakkenberg, A.; Hauer, J.; Van der Knaap, B. (1971): *Mathematische Methoden in de Sociale Geografie*. Amsterdam: KNAG and SISWO.

Dieleman, F. M.; Op 't Veld, D. (1981): Quantitative Methods in Dutch Geography and Urban and Regional Planning in the Seventies: Geographical Curricula and Methods Applied in "Spatial Research." In Bennett, R. J. (Ed.): *European Progress in Spatial Analysis*. London: Pion, 146–168.

Editorial Board (1952): Naschrift van de Redactie. In *Tijdschrift voor Economische en Sociale Geografie*, 43(8/9), 188–189.

Gauthier, H. L.; Taaffe, E. J. (2002): Three 20th Century "Revolutions" in American Geography. In *Urban Geography*, 23(6), 503–527.

Gould, P.; Leinbach, T. R. (1966): An Approach to the Geographic Assignment of Hospital Services. In *Tijdschrift voor Economische en Sociale Geografie*, 57, 203–206.

Haggett, P. (1965): *Locational Analysis in Human Geography*. London: Arnold.

Hakkenberg, A. (1969): Enige Kritische Kanttekeningen bij het Verschijnsel "Social Physics." In *Tijdschrift voor Economische en Sociale Geografie*, 60(6), 375–379.

Harvey, D. (1969): *Explanation in Geography*. London: Edward Arnold.

Hauer, J. (1967): Kritische Kanttekeningen: Peter Haggett: Locational Analysis in Human Geography. In *Bulletin van het Geografisch Instituut Universiteit Utrecht*, 4, 72–74.

Hauer, J. (1971): Een Poging tot Plaatsbepaling van de Nieuwe Geografie. In *Mathematische Methoden in de Sociale Geografie*. Amsterdam: KNAG and SISWO.

Hauer, J. (1994): Leuven Revisited, van Kwantitatieve Geografie Terug Naar de Regio. In Goossens, M.; Van Hecke, E. (Eds.): *Liber Amicorum Herman Van Der Haegen: Progress of Human Geography in Europe Van Brussel tot Siebenburgen*. Leuven: Geographical Institute, 697–705.

Hauer, J.; Van der Knaap, B. (1973): *Sociale Geografie en Ruimtelijk Onderzoek: Kwantitatieve Methoden*. Rotterdam: Universitaire Pers.

Heinemeyer, W. F.; Van Hulten, M.; De Vries Reilingh, H. D. (Eds.) (1967): *Urban Core and Inner City*. Leiden: Brill.

Heslinga, M. W. (1983): Between French and German Geography: In Search for the Origins of the Utrecht School. In *Tijdschrift voor Economische en Sociale Geografie*, 74(5), 317–334.

Het Parool (1967): Dr J. Ch. W Verstege nu aan Top van Statistiek, "CBS Geen Pure Cijferfabriek". In *Het Parool*, 4 April: 15.

Johnston, R. J. (1966): An Index of Accessibility and Its Use in the Study of Bus Services and Settlement Patterns. In *Tijdschrift voor Economische en Sociale Geografie*, 57, 33–38.

Keuning, H. J. (1948): Proeve van een Economische Hiërarchie van de Nederlandse Steden. In *Tijdschrift voor Economische en Sociale Geografie*, 39(7/8), 566–581.

King, L. J. (1962): A Quantitative Expression of the Pattern of Urban Settlements in Selected Areas of the United States. In *Tijdschrift voor Economische en Sociale Geografie*, 53(1), 1–7.

King, L. J. (1969): *Statistical Analysis in Geography*. Englewood Cliffs, NJ: Prentice-Hall.

Klaassen, L. H. (1952): De Methode van het Sociaal-economisch en Sociografisch Onderzoek. In *Tijdschrift voor Economische en Sociale Geografie*, 43(8/9), 185–188.

Klaassen, L. H.; Van Dongen Torman, D. H.; Koyck, L. M. (1949): *Hoofdlijnen van de Sociaal-Economische Ontwikkeling der Gemeente Amersfoort van 1900–1970*. Leiden: Stenfert-Kroese.

Knippenberg, H. (2008): The Transformation of the Statistical Mind in (Human) Geography. In Stamhuis, I. H.; Klep, P. M. M.; van Maarseveen, J. G. S. J. (Eds.): *The Statistical Mind in Modern Society: The Netherlands -, Volume II Statistics and Scientific Work*. Amsterdam: Aksant, 39–65.

Kouwe, P. J. W. (1985): SISWO Bureau 1960–1970. In *Berichten Over 25 Jaar SISWO*. Amsterdam: SISWO.

Kouwe, P. J. W. (1988): Sociaal Geografen en Wisseling van Context. In Van der Smagt, A. G. M.; Hendriks, P. H. J. (Eds.): *Methoden op een Keerpunt. Liber Amicorum P.J.W. Kouwe.* Utrecht: KNAG, 165–174.

Kouwenhoven, A. O. (1984): Een Ouderwetse Geograaf: Gesprekken met Marcus Willem Heslinga. In Kouwenhoven, A. O.; De Bruijne, G. A.; Hoekveld, G. A. (Eds.): *Geplaatst in de Tijd. Liber Amicorum M. W. Heslinga.* Amsterdam: Vrije Universiteit, 2–76.

Kruijt, J. P. (1944): Odium Geographicum. In *Mens en Maatschappij*, 20(2), 65–77.

Lambooy, J. G. (1966): Het Begrip Regio in de Geografische Theorie en Methode. In *Tijdschrift van het Koninklijk Nederlandsch Aardrijkskundig Genootschap*, 83, 15–23.

Lambooy, J. G. (1992): Van Economische Aardrijkskunde en Locatietheorie tot Regionale Economie. In Fase, M. M. G.; van der Zijpp, I. (Eds.): *Samenleving en Economie in de 20e Eeuw.* Leiden: Steinfert Kroese, 489–503.

Lazarsfeld, P. A.; Rosenberg, M. (Eds.) (1955): *The Language of Social Research: A Reader in the Methodology of Social Research.* Glencoe: Free Press.

Lorimer, H.; Spedding, N. (2002): Excavating Geography's Hidden Spaces. In *Area*, 34(3), 294–302.

Morrill, R. L. (1987): A Theoretical Imperative. In *Annals of the Association of American Geographers*, 77(4), 535–541.

Saey, P. (1968): A New Orientation in Geography. In *Bulletin De La Société Belge d'Etudes Géographiques/Tijdschrift Van De Belgische Vereniging Voor Aardrijkskundige Studies*, 37(1), 123–190.

Schuurmans, F.; De Vries, J. (1996): Willem Jan van den Bremen, Aardrijkskundige. In Pellenbarg, P. H.; Schuurmans, F.; De Vries, J. (Eds.): *Reisgenoten. Liber Amicorum Prof dr W.J van den Bremen.* Utrecht: KNAG, 17–56.

Schuyt, K.; Taverne, E. (2004): *1950 Prosperity and Welfare, Dutch Culture in a European Perspective.* London: Palgrave MacMillan.

Steigenga, W. (1950): Book Review of Klaassen et al. 1949, Hoofdlijnen van de Sociaal-Economische Ontwikkeling der Gemeente Amersfoort van 1900–1970. In *Tijdschrift voor Economische en Sociale Geografie*, 41(6), 153–154.

Steigenga, W. (1957): Planologisch Onderzoek 1956: Een Poging tot een Verder Program. In Hofstee, E. W. (Ed.): *Sociaal-Wetenschappelijke Verkenningen.* Assen: Van Gorcum, 177–187.

Steigenga, W. (1958): De Decentralisatie van de Nederlandse Industrie. In *Tijdschrift voor Economische en Sociale Geografie*, 49(6), 129–148.

Steigenga, W. (1964): In Memoriam. Prof dr. Johan Winsemius. In *Tijdschrift voor Economische en Sociale Geografie*, 65(10–11), 209.

Stolzenburg, R. (1984): *Het Sociaal-wetenschappelijk Onderzoek bij het Uitbreidings- en Structuurplan: Een Sociografische Opgave.* PhD Thesis. Amsterdam: Universiteit van Amsterdam.

Takes, Ch. A. P. (1948): *Bevolkingscentra in het Oude en het Nieuwe Land.* Alphen aan de Rijn: Samson.

Van Aartsen, J. P. (1953): Wiskundige Methoden bij Sociaal-geografisch Onderzoek. In *Tijdschrift voor Economische en Sociale Geografie*, 44, 50–52.

Van der Knaap, B. (1971): Een Indeling van Nederland naar "Economische Gezondheid". In *Tijdschrift voor Economische en Sociale Geografie*, 62, 332–350.

Van der Valk, A. (1983): *Opleiding in Opbouw.* Amsterdam: UvA Planologisch en Demografisch Instituut.

Van der Valk, A. (1990): *Het Levenswerk van Th.K. van Lohuizen 1890–1956.* Delft: Delftse Universitaire Pers.

Van der Wusten, H. (2004): Beyond Route 66: Some Observations on US Geography at AAG 100. In *GeoJournal*, 59(1), 47–50.

Van Doorn, J. A. A. (1956): The Development of Sociology and Social Research in the Netherlands. In *Mens en Maatschappij*, 31, 189–264.

Van Ginkel, H. (1994): Verkenner en Voortrekker: Marc de Smidt en de Utrechtse Geografie. In Atzema, O. A. L. C.; Wever, E. (Eds.): *Economisch Geografische Variaties, de Veelzijdigheid van Marc de Smidt*. Assen: Van Gorcum.

Van Ginkel, H.; De Pater, B.; Ottens, H. (Eds.) (2020): *Een Grote Stap: Het Geografisch Insituut Utrecht, 1960–1990*. Utrecht: Faculteit Geowetenschappen, Utrecht Universitity.

Van Hoof, J. C.; De Pater, B. (1982): Contouren van de Moderne Sociale Geografie. In De Pater, B.; Sint, M. (Eds.): *Rondgang door de Sociale Geografie*. Groningen: Wolters-Noordhoff, 25–38.

Van Lohuizen, Th. K.; Delfgaauw, G. Th. J. (1935): De Toekomstige Verdeeling van Bevolking en van Woningvoorraad over het Land. In *Tijdschrift voor Volkshuisvesting en Stedebouw*, 16(12), 203–209.

Van Meeteren, M. (2019a): Statistics Do Sweat: Situated Messiness and Spatial Science. In *Transactions of the Institute of British Geographers*, 44(3), 454–457.

Van Meeteren, M. (2019b): On Geography's Skewed Transnationalization, Anglophone Hegemony, and Qualified Optimism toward an Engaged Pluralist Future: A Reply to Hassink, Gong and Marques. In *International Journal of Urban Sciences*, 23(2), 181–190.

Van Meeteren, M. (2019c): The Pedagogy of Autobiography in the History of Geographic thought. In *Norsk Geografisk Tidsskrift – Norwegian Journal of Geography*, 73(4), 250–255.

Van Meeteren, M. (2022): The Polycentric Urban Region's Prehistory: A Theoretical Excavation of Mid-20th Century Dutch Applied Geography. In *Regional Studies*, 56(1), 7–20.

Van Meeteren, M. (forthcoming): Writing Blue Notes in the March of Geographical History: Revisiting the 1960 Lund Seminar in Urban Geography.

Van Meeteren, M.; Sidaway, J. D. (2020): History of Geography. In *International Encyclopedia of Human Geography*. Amsterdam: Elsevier, 2nd edition, volume 7, 37–44.

Van Paassen, C. (1989): Sociale Geografie aan de Rijksuniversiteit Utrecht. In Hoitink, H. (Ed.): *Portret van een Tachtigjarige*. Utrecht: Faculteit Ruimtelijke Wetenschappen Utrecht University, 133–148.

Van Westrhenen, J.; Dijkink, G. (1982): Onderwijsgeografie. In de Pater, B.; Sint, M. (Eds.): *Rondgang door de Sociale Geografie*. Groningen: Wolters-Noordhoff, 156–173.

Verstege, J. Ch. W. (1951): *Sociaal Onderzoek en Statistiek*. Utrecht: De Haan.

Winsemius, J. (1949): *Vestigingstendenzen van de Nederlandse Nijverheid, deel I*. Den Haag: Staatsdrukkerij.

Zeegers, G. H. L. (1946): De Opleiding van den Geograaf, Speciaal met het Oog op Onderzoekingswerk ten Behoeve van de Ruimtelijke Ordening. In *Tijdschrift van het Koninklijk Nederlandsch Aardrijkskundig Genootschap*, 63, 610–619.

3 Geographies of quantitative geographies in Brazil

Two versions of a revolution

Mariana Lamego

Introduction

I believe the most challenging moment in the research for those who decide to historicise any entity is to establish its beginning, especially for those engaged with non-traditionalist historiography. As Latour (1987) well warned us, there are plenty of entrance doors we can choose, each one will take us inescapably to different rooms and halls or even to new doors. And once we made a choice, each next step must be done in a careful way avoiding simple cause–effect relation between the historical facts and the human and non-human actors all deeply involved in the process. When dealing with a topic still not fully explored by historiography, the responsibility when creating a historical narrative that can be taken as a reference is even bigger, and a critical awareness must be all the time resting in the desktop, together with historical sources and documents.

To write on the reception, translation and spread of the quantitative revolution among Brazilian geographers is not an easy task, since had been for decades one of the most neglected subjects in Brazilian disciplinary history.[1] The overlapping in time with the infamous military dictatorship (1964–1985) together with the tidal wave of criticism in a renewal movement in social sciences and humanities in Brazil from the end of the 1970s helps understanding why the chapter referring to the so-called quantitative geography was swept under the heavy carpet of history. This substantial historical lack can be felt until nowadays and with very few exceptions (Lamego 2014, 2015; Reis Jr. 2006, 2009, 2017), Brazilian geographers still prefer the caricature of quantitative geography as ideologically reactionary, social careless and servant of the dictatorship instead of a more human and erratic version of how it was and what the adoption of quantitative techniques in the geography carried out in the country has meant.

During my research on the origins of quantitative geography caricature, I become more attentive to human affairs in the making of scientific knowledge and had engaged myself on renewed perspectives of science studies geographically sensible (Shapin and Schaffer 1985; Latour 1987; Pickering 1992; Shapin 1998). Seminal works from historical geographers firmly attached to those perspectives such as Trevor Barnes (1998, 2001a, 2001b, 2003a, 2003b, 2004a, 2004b) and David Livingstone (1995, 2003, 2004, 2007) embedded by perspectives of

DOI: 10.4324/9781003122104-3

science studies raised fundamental questions on where ideas emerged, how they emerged and what ideas emerged, putting the necessary light on the geographies of geographical knowledge. Informed by those works, I could find and tell an alternative narrative of the arrival and translation of the quantitative revolution in Brazil. A narrative that sought for personal trajectories, discovering affinity circles and, sometimes, animosities, revealing the role played by machines and scientific objects and stressing the importance and active role of place in the daily and collective work of geographers who engaged in quantitative techniques.

Some of this history will be discussed in this chapter, but, as the chosen title reveals, it will be engaged in a place-based narrative trying to cover a broad materialised network of bodies and artefacts responsible for the quantitative revolution diffusion in Brazil. The primary purpose here is to produce a geographical narrative on the uprising of a new geographical practical based on quantitative methods in two different places in Brazil in the late 1960s exploring their connections with the Anglo-American quantitative revolution. The quantitative revolution had erupted in particular places in the United States (Barnes 2004b) which helped to shape its styles of reasoning. The quantitative revolution had not remained within their birthplaces in North America, being spread for all over the country, during the 1960s, and in a short time, national borders were also crossed. The quantitative revolution reached Brazilian geography by the late 1960s. As I would like to show here, the very geography of quantitative revolution is not expressed solely by its spatial roots and character but also by its ability to travel from one place to another as books inside the luggage, as ideas insides the heads, as technical devices inside the machines. That is why my focus will be in a tangled international network through which many human and non-human actors have circulated during the mid of the 1960s until the mid of the 1970s making possible the reception and translation of quantitative revolution of geography in Brazil. The Brazilian quantitative geography[2] was forged and shaped in two different places.

The chapter has two sections. In the first one, drawing on Barnes and Abrahamsson (2017) paper, it is presented the two spots in Brazil where quantitative novelty had enrolled supporters. A brief account of the role of these places in the reception and translation of the quantitative revolution is made to understand the particular condition of the eruption of two different versions of quantitative geography in Brazil. The next section explores more closely the connections and interactions of Rio de Janeiro and Rio Claro, the two epicentres of quantitative geography in Brazil. The focus is "the corporeal travel of people" (Urry 2007, 47) revealing the existence of the international quantitative revolution network embedded in the geographies of quantitative geography in Brazil.

Expanding Peter Taylor (1977) and Barnes and Abrahamsson (2017) maps: the introduction of Brazil in the Quantgeog airlines flight plan

In 2004, Trevor Barnes presented "an interpretative analysis of the place of geography's quantitative revolution" informed by geographically sensible approaches from science studies to understand "the nature and persistence of intellectual breaks and ruptures; the embodiedness and material embeddedness of the intellectual process; and the centrality of networks and alliances" (Barnes 2004b, 565). In the text, Barnes reproduced a diagram drawn by Peter Taylor in his 1977 seminal book *Quantitative methods in geography: an introduction to spatial analysis* (Figure 3.1). The diagram shows places where the quantitative revolution first emerged, then spread to other places, and the connections between those places represented by the lines connecting the dots. Having Taylor's diagram as the base for his place-based narrative, Barnes (2004b) seeks to understand why those places, well known as "tied centres of geographical calculation central to the quantitative revolution" (Barnes 2004b, 566), were in the map and, reversely, why some other places were not. In commenting the "limited geographical reach" of Taylor's map, Barnes asks, "where is Africa, or Asia, or South America, or Australasia?" (Barnes 2004b, 566). For Barnes, "the quantitative revolution was punctured, spotty and peculiar to particular sites" (Barnes 2004b, 566) which make a materialised, embodied, situated and positional perspective of science so necessary. Adopting an 'anti-rationalist position in which place figures large as an integral component of intellectual production', Barnes (2004b) creates a narrative that binds up ideas to the heads they came from to the places where they were conceived, from the "Washington graduate diaspora" which spread the spatial cadets to different departments throughout the USA until the Lund IGU conference at 1960.

Taylor's diagram was presented again in 2017 Barnes text, co-authored with Carl Christian Abrahamsson, now introduced as "a brilliant piece of cartography because it was a map of a disciplinary idea: geography's quantitative revolution" (Barnes and Abrahamsson 2017, 105). In order to create a genealogy of spatial analysis in a people-centred account stressing the role of academic mobility and migration for the international transfer of ideas, Taylor's diagram is brought back in an expanded version with the addition of two more places linked with the previous diagram by a dotted line (Figure 3.2). The expanded map of "geography's

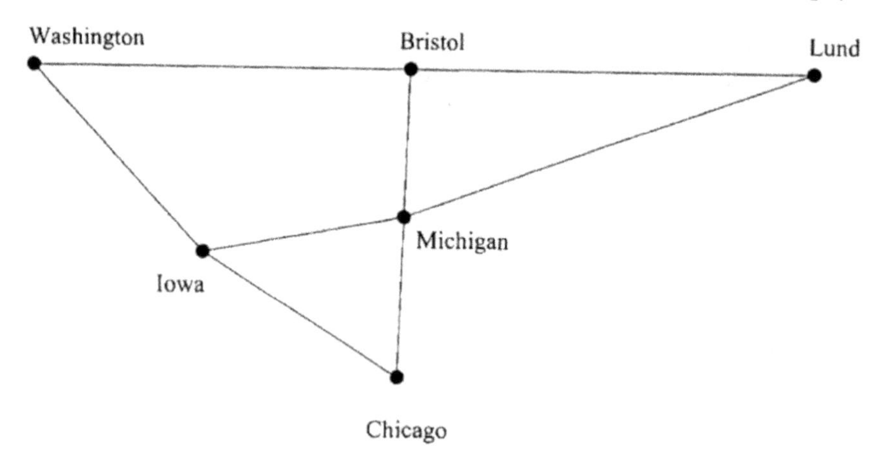

Figure 3.1 Quantgeog airlines flight plan.

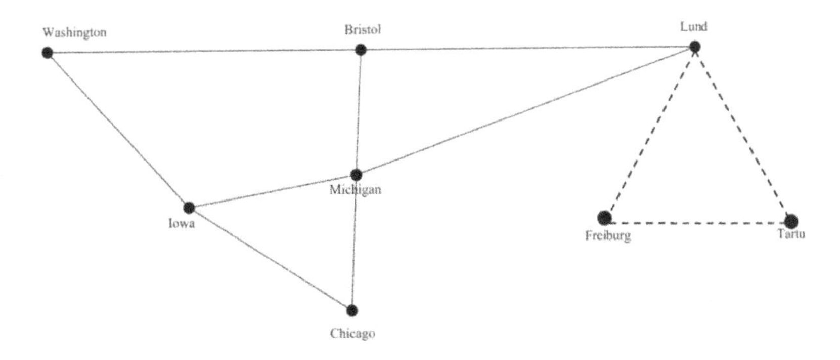

Figure 3.2 Quantgeog airlines flight plan expanded.

intellectual culture" (Barnes and Abrahamsson 2017, 106) now goes further up of the well-known centres of geographical calculation and goes back in time, showing places where the spatial analysis was previously developed.

By adding Tartu, the university capital of Estonia and Freiburg, located in southwest Germany, and creating a triangle-shaped connection between both and Lund, in Sweden, Barnes and Abrahamsson (2017) explore a prehistory of Anglo-American spatial analysis proposing "an intellectual palimpsest" with the juxtapositions of many different layers of bodies and artefacts belonging to diverse temporal and geographical contexts.

Following Barnes and Abrahamsson (2017) diagram addition, I would like to extend it one more time, incorporating in it the South American continent (Figure 3.3). As we can see in this last expanded version of Taylor's map, there is the addition of three new dots and three new connecting lines: Nottingham in the UK, Rio Claro and Rio de Janeiro, both in Brazil; and lines connecting Rio Claro to Iowa, Rio Claro to Rio de Janeiro, Rio de Janeiro to Chicago and Rio de Janeiro to Nottingham. Adopting the expedient of Barnes and Abrahamsson (2017) in designing the dotted line to represent a previous connection in the map, in the expanded version presented here, the double line was chosen to represent later connections. As a good cartographic device should be, the map informs us not only the geographical length of the quantitative revolution but also its enduring temporalities.

The enlargement of the original Taylor's diagram of the Quantgeog airlines flight plan attests the geographical mobility of quantitative revolution and its capacity of dissemination in highly different historical, geographical, cultural and disciplinary contexts. Moreover, the incorporation of Rio de Janeiro and Rio Claro in Brazil to the Quantgeog airlines flight plan raises similar questions such as the one made by Barnes and Abrahamsson (2017, 106): Why were those places on this expanded map and not others? What can one find in Rio de Janeiro and Rio Claro that made them suitable for the reception of quantitative revolution, that cannot be found elsewhere?

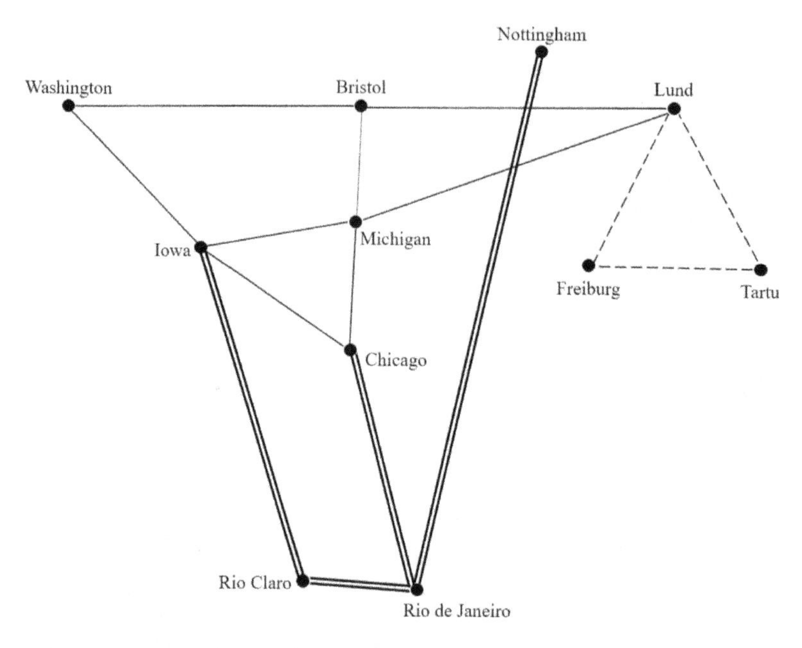

Figure 3.3 Quantgeog airlines flight plan expanded n. 2.

Rio de Janeiro and Rio Claro are the places where the quantitative revolution in Brazilian geography found great soil to be grounded. More geographically precise, the movement had its epicentres in two venues starting in the late 1960s to the mid-1970s: the Brazilian Institute of Geography and Statistics (IBGE) in Rio de Janeiro and the Faculty of Philosophy, Science and Letters of Rio Claro.

Although geographically close, both venues have some remarkable differences on a scientific, institutional and political level, which help sustain that different regional styles of reasoning were straightly related with a diverse institutional arrangement and a diverse interpretive community in which quantitative claims were received.

IBGE is located in Rio de Janeiro city, a big metropolis economically and politically influential for the whole country. IBGE is not an academic institution, nor a scientific society. It is a government agency, created during the 1930s to provide and systematise social, economic and territorial information of the country in a period of centralisation and rationalisation of state politics known as Estado Novo (1937–1945) under the leadership of President Getúlio Vargas. The Geography National Council (CNG) works within the IBGE structure, and since the beginning of its operation in 1938, it counted on the participation of renowned foreign geographers, from France, UK and USA, who worked in the formation of the technical staff of the council. From the late 1940s until the mid-1970s, IBGE became central in urban and regional planning politics (Corrêa 1980; Penha 1993; Almeida 2000; Bomfim 2007).

The Faculty of Philosophy, Science and Humanities of Rio Claro is located in Rio Claro, a small and provincial town at São Paulo state. It is a public state university, established in 1958, due to a state politics to expand higher education institutions for the countryside. The first professors of the Department of Geography at Rio Claro came borrowed from central and reference national institutions such as the University of São Paulo and from the CNG from IBGE.

These geographical contingencies help explaining differences between the content and practices of quantitative geography of IBGE from that of Rio Claro University. However, even though place contingencies are crucial, they are not sufficient once it is necessary to consider the role played by the geographical mobility of knowledge. That is why to understand the reception and translation of quantitative revolution in Brazilian geography, we must drive our eyes far beyond the local. After all, the expanded map is not made only by dots. We shall consider the lines between the dots, raising question on how the connections and interactions with other venues also made possible the transfer, adaption and transformation of knowledge and ideas of quantitative revolution. The relation between the dots and the lines, i.e. the local contingencies shaping knowledge and the transformation made possible by the knowledge mobility is dialectical in the case of the quantitative revolution in Brazilian geography. If the place shapes knowledge, it is because the trip makes it suitable. Nevertheless, the existence and the nature of an international network also shape local contingencies, both in Rio de Janeiro and Rio Claro.

Quantitative revolution approaches Rio de Janeiro and Rio Claro: early connections and interactions

Looking at the expanded map in Figure 3.3, Rio de Janeiro connections exceed in amount when compared to Rio Claro. The explanation came from the local contingencies explored earlier, from the fact that as a government agency, IBGE easily fits in the Latour (1987) notion of centre of calculation, well employed in Barnes (2004b) analysis on the quantitative revolution in particular places. Until the early 1980s – when the agency lost its importance due to the creation of other strategical government agencies – IBGE was a place that developed strategies and methods for collection, accumulation, organisation and dissemination of information. IBGE has been a node in an international network of alliances on geographical theories and practices implicated at national policy issue of urban or regional planning mainly (Barnes 2004b, 577). The connections of Rio de Janeiro in the Quantgeog airlines plan during the end of the 1960s were made possible by previous connections that IBGE have had with the USA and UK geographies institutions.

Being part of an international network of geographical practices and ideas made the CNG inside IBGE a very porous place. Since its creation in 1934, the CNG received visitors from many countries, geographers, and geomorphologists, especially from France, German, the USA and the UK. Looking back in the institutional history, it is possible to notice some shifting regarding the nationalities of the visitors according to the period which is in its turn directly related to the geography practised at the IBGE and also to the Brazilian governmental spatial politics. The first and most lasting relationship is undoubted with French geographers. From the 1930s to the 1960s distinguished French geographers, such as Pierre Deffontaines, Francis Ruellan, Jean Tricart, Michel Rochefort, among many others, with their busy head ideas and their luggage full of books arrived in Rio, and were largely responsible for the implementation and consolidation of scientific investigations and academic institutions in our country,[3] technical training courses in geography,[4] as well as the development of huge planning projects at IBGE,[5] as part of its internationalisation policy (Machado 2000). And once the path is cleared, it becomes two-way soon. The other part of the IBGE internationalisation policy was to send its employers to have professional or academic qualification courses.[6]

During the Second World War, the connections between French and Brazil were weakened enabling the set-up of new connections with the USA,[7] especially through the arrival of Brazilian geographers from IBGE for Master and PhD degrees at Wisconsin and Chicago mainly.[8] After the war, this new interaction circuit of IBGE with the USA had as a consequence the opening of new investigation ways in Brazilian geography from IBGE, with the implementation of regional studies, fieldwork methodologies and colonisation process (Almeida 2000, 88). Brazilian geographers normally familiar with French geographical theories and methods started an orientation towards a distinct geographical knowledge-making and knowledge-communicating as well.

One of the earlier visitors from the USA to IBGE was the North American geographer and professor at Syracuse University Preston E. James who came to Brazil for a year of research from 1951 to 1952. Preston E. James, who during the wartime was Chief of the Latin American Division of the Office of Strategic Services (Jensen 1986), was assisted by Speridião Faissol, an IBGE geographer who has, among his skills speaking fluent English, a still unusual proficiency among Brazilian geographers much more familiar with French. The assistance enables Faissol to earn a scholarship for a PhD study under the supervision of Preston James, at that time head of the Geography Department at Syracuse University from 1952 to 1956.

From the end of the 1960s, the echoes of the quantitative revolution in the Anglo-American world could be heard at IBGE relatively quickly because of Faissol's role, who had become the main interlocutor regarding the reception of Anglo-American world researchers. Speridião Faissol was a great enthusiast for the adoption of new techniques and methods in the research carried out by IBGE geographers, especially in the theme of urban and regional planning, previously influenced by Michel Rochefort's theory of development poles. As his power in the hierarchy of IBGE's professional staff increased – Faissol became the director of the Department of Geography and became general secretary of the National Council of Geography – Faissol could take a substantial turn towards quantitative geography, recruiting a select team of IBGE professionals interested in applying sophisticated mathematics in their research.

Thanks to some of the early connections made, as mentioned above, the Rio de Janeiro connections lines were already constituted when John Peter Cole from Nottingham and Brian Berry from Chicago have taken their plane to Rio de Janeiro between 1967 and 1969. Professor John P. Cole firstly went to Brazil in 1967 due to a scholarship he earned from the British government to study the Brazilian urban system. According to an interview with Faissol (1997), Cole's arrival was published in a newspaper of the time. Faissol had read the news and got in touch with the English researcher inviting him to visit IBGE (Faissol 1997). Cole and Faissol became very close friends, and during the following year, Cole returns many times to Brazil officially to training courses for IBGE geographers who later would integrate the GAM, the Group of Metropolitan Areas, created in 1969 by Speridião Faissol. The GAM had become the quantitative revolution's headquarters in Rio de Janeiro. According to Faissol (1997), Cole taught factorial analysis with all data previously sent to Nottingham, because at that time IBGE did not have the appropriate computer software to apply the quantitative methods. This situation only changed in 1971, when IBGE had received a new generation of IBM computers primarily to prepare the Census.

The partnership between Cole and Faissol was fundamental in those first years of the constitution of Brazilian quantitative geography. Cole has assisted the GAM in introducing quantitative techniques such as factor analysis and clustering methods, Markov Chain method, among others. The opened connection Rio de Janeiro-Nottingham was used by Brazilian geographers from the GAM such as Olga Maria Buarque de Lima, who went to the University of Nottingham for her master studies under Cole's supervision in 1973.

The importance and influence of Cole in the researches of IBGE geographers can be felt in several papers presenting results of main research carried out at the institute and published at the main vehicle of diffusion of the production of IBGE, the journal *Revista Brasileira de Geografia* (RBG).[9] It is the case of the paper signed by Pedro Geiger, a well-known Brazilian geographer who at the end of the 1960s, got deeply engaged in quantitative methods in his research. The paper was published in 1971, in the special issue of RBG entirely dedicated to present researches made at IBGE using quantitative methods, and presents the application of factor analysis methods in the study of cities, in this case, northeastern Brazilian cities. In the introduction to the article, Geiger mentions that the "computation operations, from the first composite matrix . . ., were carried out at the University of Nottingham, England, kindly of one of his professors from the Department of Geography, our friend John Cole" (Geiger 1970, 132). In the same special issue, there is a paper by Cole, together with Faissol and McCullagh in which it was employed the Markov Chain methods to project the growth of the Brazilian population and the internal migrations (Faissol et al. 1970).

Regarding the visit of Brian Berry, not so long as Cole's visit but also very significant in the later development of quantitative geography in Brazil, it was Speridião Faissol again who, in 1969, received Berry and the economist John Friedmann for a series of urban planning classes to Brazilian geographers and economists from IBGE. The history behind Berry and Friedmann arrival in IBGE is quite odd and expresses how hints of chance can join the will in the making of knowledge. According to Faissol (1997), Berry and Friedmann were originally invited by the Brazilian urbanist and architect Harry Cole for consultancy services at the SERFHAU (State Agency of Social Housing and Urbanism). It turns out that on the eve of the arrival of Berry and Friedmann, Harry Cole got fired from his post. Cole was helpless when he called Faissol in the middle of the night offering, according to Faissol's words (1997), the two Americans to his friend. With no hesitation, Faissol accepted the offer and had received both Berry and Friedmann for a series of lectures and courses with geographers from Faissol personal staff at IBGE.[10] Berry at that time was already known in Brazil because of his engagement on the quantitative revolution in Chicago.[11] Berry's visit made him very close to some Brazilian geographers from IBGE, who had the opportunity to go to Chicago for PhD and Masters studies such as Roberto Lobato Corrêa (Almeida 2000).

The influence of Berry's course at IBGE can be felt in several papers made by IBGE geographers in which some of the techniques taught are applied, such as Fany Davidovich's (1969) paper on flow studies as a subsidy to regionalisation at the RBG, in which Davidovich applied the intervening opportunity model to solve some problems related to the immense volume of data in the study of flows.

Berry's relationship with IBGE appears materialised in the pages of the special issue of the RBG on quantitative methods. The article signed by Brian Berry and Gerald Pyle (1979)[12] consists of an "additional quantitative regionalisation essay" (Berry and Pyle 1979, 30). The objective was to produce, applying quantitative techniques and factor analysis, a regional division that, following the criteria used by IBGE geographers, could be compared with that developed by IBGE in a traditional methodology. Once compared, the authors analyse their similarities and differences. It is impossible not to perceive this "essay" as an element of a convincing strategy in favour of the use of quantitative techniques, especially in a subject whose data volume is challenging to manage. For the study, made in 1968, IBGE research was the source of all data, and the analysis using quantitative methods and factor analysis, as well as the interpretation of the results, were done at the University of Chicago, where Berry was based. As in the same case of Cole's participation in the development of quantitative methods in the geography of IBGE, the processing of data could not be made in Brazil, since the needed technologies were not available at that time.

From 1970 onwards, the already consolidated network was enlarged by the meeting of the IGU Commission on quantitative methods in April 1971, at National School of Statistical Sciences[13] facilities. The meeting is regarded as one of the highlighted moments in the history of quantitative revolution in

Brazil, stressing the leading prominence of IBGE geography in that context. The meeting "aimed to disseminate in Brazil, at the initiative of The Geography Department at IBGE and interested institutions, the use of Quantitative Methods in the Analysis of Brazilian Geographical Problems" (IBGE 1971, 1). Part of the papers presented at the meeting was published in IBGE's Brazilian Journal of Geography. It was during this time that Faissol became secretary of the commission by Brian Berry nomination (Faissol 1997).

The meeting had the presence of the professors from Rio Claro engaged in quantitative methodology, which explains the connection between Rio de Janeiro and Rio Claro in the expanded version of the Quantgeog flight plan. The reception of the quantitative revolution among geographers from the Department of Geography at Rio Claro University was quite different when compared to IBGE geographers bringing forth a different version of quantitative geography. If it is possible to suggest that IBGE functions as a calculation centre, according to Latour's conception, in regarding the Faculty of Philosophy and Letters of Rio Claro, it is more appropriate the notion of heterotopia in explaining how, in a provincial city beyond the centres of geographic knowledge production borders in the country, the quantitative revolution could find a fertile ground to blossom.

As developed by Barnes (2004b), heterotopia would be a specific condition that certain places have of not conforming to surrounding norms or standards. Drawing by Hetherington (1997) formulations, "heterotopia are spaces in which an alternative social ordering is performed . . . in which a new way of ordering emerges that stands in contrast to the taken-for-granted mundane idea of social order that exists within society" (Hetherington 1997, 40). Regarding the place-based perspective of knowledge production, heterotopia would be places of potential intellectual innovation, those who suspect, neutralise, or even reverse the set of relationships that supposedly should mirror.

The small group of professors from the Department of Geography of Rio Claro seemed to have met the requirements for the place of potential intellectual innovation. They were mostly young, in their first decade as university professors. They were not under the influence of traditional French geography, in a peripheral university centre. Moreover, they were looking for new trends in discipline practice and theories. The will of Rio Claro group in embracing the quantitative revolution is related to a detached condition of Rio Claro geography concerning geography practised in the University of São Paulo, one of the largest university centres in the state as well as in the country. Opposite of the University of São Paulo, at Rio Claro Geography Department, it has never been performed traditional French geography. Despite being a few hundred kilometres from São Paulo geography, Rio Claro's geography would be methodologically much further away.

This specific condition granted relative autonomy to Rio Claro professors concerning the undergraduate curriculum, to follow a previously unexplored line in universities. In this sense, it was possible to exercise its heterotopic condition. Professors from Rio Claro could then experience the juxtaposition of themes and methods never practised before, not at a Brazilian university.

The arrival of the quantitative revolution among the Rio Claro group can be traced and represents the connection between Rio Claro and Iowa in the expanded version of the Quantgeog airlines plan (Figure 3.3). The Brazilian professor Lívia de Oliveira plays a distinctive role in this history. In 1968, after attending the 21st International Geographical Congress in India, Lívia de Oliveira, professor at Rio Claro Department of Geography, had heard from a couple of geographers from South Korea that Iowa City had become a high centre of intellectual effervescence (Oliveira 2007). Lívia decided to visit Iowa, and during her one month stay, she made frequent visits at the University of Iowa, where she met some professors from the Department of Geography who had confirmed that the Department had long abandoned mere descriptions, adopting modern statistical tools instead. Fascinated by the collections of the local University Library, Lívia returned to Brazil with additional baggage: works proving the scientific changes as well as information on key figures.

Partly nourished by Lívia's precious baggage,[14] and partly due to an evident dissatisfaction with the geography that was being done in the department, in 1969 the most prominent figure in the Rio Claro group, Professor Antonio Christofoletti had started a study group on the Quantitative Revolution in the USA with other colleagues and PhD students such as Antonio de Olívio Ceron, José Paulo Filizola Diniz, Maria Lúcia Helena Gerardi, among many others. The mood for the study group meetings was the possibility of debating geography novelty from abroad. The study group's stimulus was, above all, the prospect that a methodological renewal never seen before in the discipline would lead, according to Diniz (2004) "geographers to replace their basic interest of pure location and description of facts with a concern to identify and explain spatial structures and processes".

The meeting group had got substance and in 1971, it was founded that the Rio Claro Association of Theoretical Geography (AGETEO) was reflecting the particular interest of its founding geographers in spreading and consolidating a new practice in the discipline. Convinced that they were facing an unprecedented revolution, capable of erasing the traditional dichotomies that afflicted geography, they set out to create the Theoretical Geography Bulletin (BGT), whose purpose was to "facilitate access to new ideas arising from the 'methodological renewal' that occurred in North American geography" (vol. 1, p. 1). The BGT has shortly become the diffusion vehicle of growing domestic production of quantitative geography as well as a translation space of classical text on quantitative geography from abroad.

Conclusion

The history of the reception, adoption and translation of Anglo-American quantitative revolution in Brazil can be told through many entrance doors. Aware of Latour's (1987) warning, I presented the history of quantitative geography in Brazil based on personal networks and the mobility of people and objects. Geographical mobility played a central role in the production and dissemination of knowledge, and the histories of the quantitative revolution in Brazil could show that local contingencies are fundamental to understand how knowledge was shaped by varying social, cultural, economic and political contexts (Jöns et al. 2017, 12).

The inclusion of Rio de Janeiro and Rio Claro in Taylor's Quantgeog airlines flight plan (Barnes 2004b; Barnes and Abrahamsson 2017) demonstrate how places could be understood considering their connections with other places. When considering these connections, we have to insert other places not included in previous networks, such as Nottingham, which was crucial in IBGE networks due to the mobility of John Peter Cole. Chicago also appears as an essential point, through the mobility of Brian Berry to Brazil. And these were not one-way paths. During the quantitative revolution, some Brazilian geographers went to Nottingham and Chicago in search of their graduate studies. With Professor Faissol as an important actor in a network of international collaborators, IBGE became a place for the application of methods, techniques and practices associated with quantitative geography.

If the arrival of Anglo-American geographers reinforced IBGE as a centre of calculation, in the case of Rio Claro, it was the isolation and mobility of specific actors abroad that guaranteed its heterotopic condition. Besides Rio Claro local contingences gathering geographers from São Paulo University and IBGE, unsatisfied with traditional regional approaches applied in these places, the mobilities of Professor Lívia de Oliveira to Iowa and the books she brought in her luggage were central to understand self-nomination as "theoretical" by local geographers. These geographical contingencies help explaining differences between the content and practices of quantitative geography of IBGE from that of Rio Claro University, and these differences and networks were materialised in the journals published by these institutions, *Revista Brasileira de Geografia* and the *Boletim de Geografia Teorética.*

In addition to people and books, the mobility of machines is another crucial element to understand different translations of the quantitative revolution in Brazil. Before the arrival of a new generation of computers to carry out the 1971 census, IBGE had to send its data to be processed in other centres of calculation, such as Nottingham and Chicago. The acquisition of these machines allowed the implementation of software for quantitative methods and finally gave the institution the capacity to apply new research techniques. In the case of Rio Claro, the absence of non-human artefacts, such as modern computers, had strengthened the connections between Rio Claro and Rio de Janeiro, since the quantitative analyses were carried out on IBGE computers.

Considering how the mobility of people and objects shapes the local conditions of knowledge production, the rooms and halls covered in this work lead us to new doors. Following Barnes' call for a more comprehensive Quantgeog airlines flight plan in time and space, many places can be inserted in future research, showing how the production of a global history of the quantitative revolution needs to take into account many local translations and particular place-based contingencies.

Notes

1 Only very recently we can see the increase of interest among a new generation of researchers, being this present volume with the contributions of Lira and Ribeiro, an excellent example of this new trend among Brazilian scholars.
2 The expressions "quantitative revolution", "spatial analysis" or "spatial science" did not have the same impact in Brazil as in the USA or the UK. In Brazil, three nominations have been adopted more frequently in the following order: *geografia quantitativa* (quantitative geography); *geografia teorética quantitativa* (theoretical quantitative geography) and *geografia neopositivista* (neopositivist geography).
3 On the French mission in developing the first modern universities in country, see Larissa Alves de Lira's chapter in this volume.
4 In the same year of 1935, Deffontaines went to Rio de Janeiro helping with the implementation of the university, the French geographer also started technical training courses in geography with the first generation of geographers from IBGE (Almeida 2000).
5 French geographer Michel Rochefort went to Brazil for the first time in 1956, on the occasion of the 18th International Geographical Congress of International Geographical Union. It was the beginning of a long relationship between Rochefort and Brazilian geographers. Extending his stay, in the early 1960s, Rochefort starts a partnership with the CNG of IBGE, responsible for coordinating three major research projects: Human Potential, Industrial Geography and Urban Geography (Bomfim 2015).
6 The Second World War postponed the sending of Brazilian geographers from IBGE to PhD and Master degrees in French Universities. Just after the war, in 1946, Miguel Alves de Lima, Pedro Geiger, Elza Keller, Miriam Mesquita and Héldio Xavier César went to France indicated by Francis Ruellan, who worked in Brazilian universities and at IBGE, appointed as Technical Assistant of the National Council of Geography, from 1940 to 1956. Ruellan was responsible for the training of the second generation of Brazilian geographers (Machado 2000).
7 It is important to understand this IBGE shift towards the USA as part of a major strategy of Brazilian foreign policy within the pan-Americanism movement.
8 According to Almeida (2000) from 1942 until 1945, the first generation of geographers at IBGE went to the USA to have Master and PhD degrees. Jorge Zarur went to Wisconsin in 1942 for his Master degree and in 1943 to Chicago for a specialisation fieldwork course. In 1945, it was the turn of Fábio de Macedo Soares Guimarães and Orlando Valverde who went to the University of Wisconsin, Lúcio de Castro Soares and Lindalvo Bezerra dos Santos who went to the University of Chicago and José Veríssimo da Costa Pereira who went to the University Northwestern (Almeida 2000, 120–121).
9 The RBG is the main vehicle of diffusion of the production of IBGE. Created in 1938 it is, until nowadays, an important and historical source of geographical knowledge production in Brazil.
10 The economist and planning theorist John Friedmann had a close relation with Brazil, since 1955 when he visited the country under a USAID (United States Agency for International Development) agreement with SPVEA (Superintendence of the Amazon Economic Enhancement Plan) in the north Brazilian region (Chiquito 2006). From 1956 to 1958 John Friedmann taught at the Federal University of Bahia.
11 According to an interview for Almeida (1995), Faissol referred to Berry as the "prince of the quantitativism" (176).
12 At that time a PhD student at University of Chicago.
13 The National School of Statistical Sciences (ENCE) was founded in the early 1950s as a federal institution of higher education dedicated to teaching statistics, associated to IBGE.
14 Among the books on quantitative revolution, Livia Oliveira (2007) highlights Bunge's Theoretical Geography (1962) and Haggett's Locational Analysis (1966).

References

Almeida, R. S. de. (2000): *A Geografia e os Seógrafos do IBGE no Período de 1938–1998*. Tese de Doutorado. Programa de Pós-Graduação em Geografia. Instituto de Geociências, UFRJ.

Almeida, R. S. de. (1995): Memória: Speridião Faissol. In *Caderno de Geociências (IBGE)*, (15), 165–182.

Barnes, T. J. (1998): A History of Regression: Actors, Networks, Machines and Numbers. In *Environment and Planning A*, 30, 203–223.

Barnes, T. J. (2001a): Lives Lived, and Lives Told: Biographies of Geography's Quantitative Revolution. In *Society and Space: Environment and Planning* D, 19, 409–429.

Barnes, T. J. (2001b): "In the Beginning Was Economic Geography": A Science Studies Approach to Disciplinary History. In *Progress in Human Geography*, 25, 455–478.

Barnes, T. J. (2003a): What's Wrong with American Regional Science? A View from Science Studies. In *Canadian Journal of Regional Science*, 26, 3–26.

Barnes, T. J. (2003b): The Place of Locational Analysis: A Selective and Interpretive History. In *Progress in Human Geography*, 27, 69–95.

Barnes, T. J. (2004a): A Paper Related to Everything, But More Related to Local Thing. In *Annals, Association of American Geographers*, 94, 278–283.

Barnes, T. J. (2004b): Placing Ideas: Genius Loci, Heterotopia, and Geography's Quantitative Revolution. In *Progress in Human Geography*, 29, 565–595.

Barnes, T. J.; Abrahamsson, C. C. (2017): The Imprecise Wanderings of a Precise Idea: The Travels of Spatial Analysis. In Jöns, H.; Meusburger, P.; Heffernan, M. (Eds.): *Mobilities of Knowledge*. Cham: Springer, 105–121.

Berry, B. J. L.; Pyle, G. F. (1979): Grandes Regiões e Tipos de Agricultura no Brasil. In *Revista Brasileira de Geografia*, 32(4), 23–40.

Bomfim, P. A. de A. (2007): *A Ostentação Estatística (um Projeto Geopolítico Para o Território Nacional: Estado e Planejamento no Período Pós-64)*. Tese de Doutorado. Programa de Pós-Graduação em Geografia. Faculdade de Filosofia, Letras e Ciências Humanas, São Paulo, USP, 377p.

Bomfim, P. A. de A. (2015): Michel Rochefort e o Instituto Brasileiro de Geografia e Estatística na Década de 1960. In *Sociedade e Natureza Uberlândia*, 27(3), 365–378.

Bunge, W. (1962): *Theoretical Geography*. Sweden: The Royal University of Lund, C.W.K Gleerup Publishers.

Chiquito, E. (2006): Entrevista com John Friedmann. In *Risco Revista de Pesquisa em Arquitetura e Urbanismo*, 14(2), 82–89.

Corrêa, R. L. (1980): Da "Nova Geografia" à "Geografia Nova". In *Revista de Cultura Vozes*, ano 74, 74(4), 253–260.

Davidovich, F. (1969): A Experiência dos Estudos de Fluxos, no IBG, como Subsídio à Regionalização. In *Revista Brasileira*, 31(2), 66–80.

Diniz, J. A. F. (2004): Entrevista com o Professor José Alexandre Filizzola Diniz. In *Geosul*, 19(37), 215–231.

Faissol, S. (1997): "Cinqüenta anos de Geografia." Entrevista com o Professor Speridião Faissol (Conduzida por Helion Povoa Neto e João Rua). In *GeoUERJ*, (1), 55–70.

Faissol, S.; Cole, J. P.; McCullagh, M. J. (1970): Projeção da População no Brasil – Aplicação do Método Cadeia de Markov. In *Revista Brasileira de Geografia*, 32(4), 173–208.

Geiger, P. (1970): Cidades do Nordeste. Aplicação de "Factor Analysis" no Estudo de Cidades Nordestinas. In *Revista Brasileira de Geografia*, 32(4), 131–172.

Haggett, P. (1966): *Locational Analysis in Human Geography*. New York: St. Martin's Press.

Hetherington, K. (1997): *The Badlands of Modernity*. London: Routledge.

IBGE (1971): "Reunião da Comissão de Métodos Quantitativos". In *Revista Brasileira de Geografia*, 33(2), 150–151.

Jensen, R. G. (1986): Memorial Preston James, 1899–1986. In *Journal of Geography*, 85(6), 273–274.

Jöns, H.; Meusburger, P.; Heffernan, M. (Eds.) (2017): *Mobilities of Knowledge*. Cham: Springer.

Lamego, M. (2014): O IBGE e a Geografia Quantitativa Brasileira: Construindo um Objeto Imaginário. In *Terra Brasilis (Nova Série)* [Online], 3. https://doi.org/10.4000/terrabrasilis.1015.

Lamego, M. (2015): Genius loci: Duas Versões da Geografia Quantitativa no Brasil. In *Terra Brasilis (Nova Série)* [Online], 5. https://doi.org/10.4000/terrabrasilis.1504.

Latour, B. (1987): *Science in Action: How to Follow Scientists and Engineers through Society*. Cambridge: Harvard University Press.

Livingstone, D. (1995): The Spaces of Knowledge: Contributions towards a Historical Geography of Science. In *Environment and Planning D: Society and Space*, 13, 5–34.

Livingstone, D. (2003): *Putting Science in Its Place*. Chicago and London: The University of Chicago Press.

Livingstone, D. (2004): Keeping Science in Site. In *Historically Speaking*, 5(3), 10–12.

Livingstone, D. (2007): Science, Site and Speech: Scientific Knowledge and the Spaces of Rhetoric. In *History of the Human Sciences*, 20(2), 71–98.

Machado, M. S. (2000): A Implantação da Geografia Universitária no Rio De Janeiro. In *Scripta Nova: Revista Electrónica de Geografía y Ciencias Sociales*, (4), 69.

Oliveira, L. (2007): Entrevista com a Professora Lívia de Oliveira. In *Geosul*, 22(43), 215–231.

Penha, E. (1993): *A. A Criação do IBGE no Contexto da Centralização Política do Estado Novo*. Rio de Janeiro: Fundação IBGE.

Pickering, A. (1992): From Science as Knowledge to Science as Practice. In Pickering, A. (Ed.): *Science as Practice and Culture*. Chicago: University of Chicago Press, 1–26.

Reis Junior, D. F. C. (2006): Valores e Circunstâncias do Pensamento Geográfico Brasileiro: a Geografia Teorética Ponderada de Speridião Faissol. In *Geografia* [En Ligne], 31(3), 481–504.

Reis Junior, D. F. C. (2009): Valores e Circunstâncias do Pensamento Geográfico Brasileiro: a Geografia Teorética Transitiva de Antonio Christofoletti. In *Geografia* [En Ligne], 34(1), 5–32.

Reis, D. F. C., Jr. (2017): A Historiography of Brazilian Theoretical and Quantitative Geography: The "Rio Claro Case", from Flourishing to Fall. In *Cybergeo: European Journal of Geography* [En Ligne], Epistémologie, Histoire de la Géographie, Didactique, Document 837.

Schwartzman, S. (1979): *Formação da Comunidade Científica no Brasil*. São Paulo: Companhia Editora Nacional e Finep.

Shapin, S. (1998): Placing the View from Nowhere: Historical and Sociological Problems in the Location of Science. In *Transactions of the Institute of British Geographers*, New Series, 23, 5–12.

Shapin, S.; Schaffer, S. (1985): *Leviathan and the Air-Pump: Hobbes, Boyle and the Experimental Life*. Oxford: Princeton University Press.

Taylor, P. J. (1977): *Quantitative Methods in Geography: An Introduction to Spatial Analysis*. Boston: Houghton Mifflin.

Urry, J. (2007): *Mobilities*. Cambridge: Polity Press.

4 Translation of quantitative geography in the Brazilian journals

The cases of the *Boletim Geográfico* (1966–1976) and *Revista Brasileira de Geografia* (1970–1982)

Guilherme Ribeiro

Introduction

From the standpoint of the history of geography, globalization has at least three effects: the internationalization of knowledge; the wide circulation of books, authors, and bulletins; and the dominance of the English language. From the position of a Brazilian geographer despite social structures and political circumstances, the scenery is positive: we have been increasing external research networks; we travel with some frequency to foreign congresses; and it is not difficult to get updates on European or Asian literature.

Conversely, there are still hidden hierarchies. I only knew that I am a sort of tropical-peripheral scholar attending "international" symposiums (see Novaes 2015; Gyuris 2018) where both my language and my literature were unknown for the participants. In contrast, I was aware of their literature and some of their most significant debates. In general, I feel that "peripheral" scholars learn more languages than "central ones." The Indian thinker Gayatri Chakravorty Spivak profoundly noted, "If you are interested in talking about the other, and/or in making a claim to be the other, it is crucial to learn other languages. This should be distinguished from the learned tradition of language acquisition for academic work" (Spivak 2000 [1993], 407). Indeed, language is an unescapable matter for those who are engaged in making a democratic dialogue among scholars from distinct nationalities and backgrounds. With this in mind, the role of translations is very relevant, which has been usually seen as a "natural" subject but they are indeed geopolitical matter. That is one of the reasons why the field of translation studies – my methodological support in this text – established a stimulating interaction with geography and, in line with the spatial turn (Bachmann-Medick 2016 [2015]), concepts such as "translating geographies" (Italiano 2012, 2016), "political geographies of displacement" (Simon 2009), and "spatial operator" (Cronin 2003, 2010) have opened up stimulating empirical and theoretical possibilities for both domains.

Hence, the purpose of this text is to examine the history of Brazilian geography through translations or, more accurately, the translations of quantitative geography published by the *Revista Brasileira de Geografia* and the *Boletim*

DOI: 10.4324/9781003122104-4

Geográfico. Both were prepared by the *Instituto Brasileiro de Geografia e Estatística* [Brazilian Institute of Geography and Statistics; hereafter, IBGE], the most expressive Brazilian geographical organization in the last century. If the writings about the Anglo-Saxon case are much known in its multiple aspects and approaches[1] (Martin 1972; Bunge 1979; Billinge et al. 1984; Barnes 2004, 2016; Goodchild 2008; Charlton 2008; Johnston 2008), the Brazilian chapter of the new geography is still restricted to the Portuguese readers. Despite the high level of Brazilian researches concerning quantitative geography (Reis Jr. 2007; Lamego 2010), this is not enough to include them in the "international" literature. That is one of the reasons why I decided to write this paper in English.

So during the period extending from 1966 to 1982, I will identify authors, sources, languages, translators, and themes in order to write an unusual chapter on how spatial analysis circulated in Brazil. My main argument is that in a peripheral country translation is an essential tool in terms of academic modernization. In this case, translating English papers into Portuguese had a crucial role toward the introduction of quantitative methods in Brazil in the end of the 1960s and, consequently, in the process of putting in question the hegemony of French geographers since the 1930s related to the classic methods such as fieldwork, regional exploration, and the scrutiny of the visible aspects of the landscape.

Albeit the two journals aforementioned have published a range of subjects, concerning translation they emphasized quantitative texts more than another in the period of 1966–1982. This is due to two reasons: first, part of Brazilian geographers adopted quantitative methods because they were convinced that this approach was the most useful in order to explain the spatial challenges of that time such as the growth of urbanization and industrialization in a Third World country. Second, we cannot forget that IBGE is a federal institute and that we experienced a military dictatorship between 1964 and 1985. So, I think that quantitative geography and its emphasis on the economic capitalist development were fitted like a glove in that context.

In the first section of this chapter, I discuss the relations between translation and history of Brazilian geography in order to reveal how translation is a political issue for a geographer from the Global South within a global circulation of knowledge, and how we can analyze it according to translation studies. In the second section, I address a quantitative and qualitative investigation with reference to translations of spatial analysis performed by *Boletim Geográfico* (1966–1976) and *Revista Brasileira de Geografia* (1970–1982). In the last section, I will join empirical data and theoretical support to further understand the translations of spatial science in Brazil.

Science, translation, and history of geography from a Brazilian point of view: issues of method and politics

Our understanding of "science" has changed since French intellectuals such as Michel Foucault and Pierre Bourdieu have examined science as a field of conflicts, which means considering how institutions, experts, and journals not only as including but also excluding concepts, methods, and subjects in order to maintain structures of power, influences, publishing, grants, prizes, and so on. In analyzing the sociopolitical dimensions of science, they have contributed to questioning one of the most important taboos from the modern thought: the impartiality of researchers and the neutrality of knowledge (Foucault 1971; Bourdieu 1976).

In transplanting this framework to the global scale, it is easy to observe the establishing of "centers of calculation" (Latour 1987) assembled in a small number of laboratories and universities in Western Europe, the United Kingdom, and the United States. These centers have attracted both young and mature scholars from different parts of the world in search of legitimization in their home countries. It looks like we are facing a unidirectional movement because, when peripheral scholars arrived at those universities, they already are assigned to a discursive order, what means to say that they know its languages, authors, and books. As a result of this situation, "hierarchies," "traditions," and "canons" are being reproduced throughout the world in a noncritical way (Keighren et al. 2012). Furthermore, an intellectual division of labor is established where centers represent the *avant-garde* of knowledge and peripheries are no more than mere consumers of ideas. In addition, many of these ideas are already outdated when they reach the margins. This is one of the reasons why the Peruvian sociologist Aníbal Quijano realized that even the imaginary has already been colonized (Quijano 2000). In the same vein, the Cameroonian philosopher Achille Mbembe points out that when a black man asks himself who he is, the reply belongs to the white man (Mbembe 2013).

Summarizing Brazilian geography in the 20th century, most of the possible answers would take into account the role of foreign influences. The first concerns the coming of the French Mission in the 1930s to found Faculties of Human Sciences in São Paulo and Rio de Janeiro (Miceli 1989; Machado 2009). For instance, most of their classes were taught in French. In an ancient colony greedy of modernity, and despite the fact that many of them were young scholars, the arrival of Pierre Deffontaines, Francis Ruellan, Pierre Monbeig, Fernand Braudel, Claude Lévi-Strauss, and others becomes a sign of a new era but also the cradle of some of the best outcomes from regional geography, long-run history, and structural anthropology in close relation to the Brazilian experience (see Delfosse 1998; Paris 1999; Moraes and Rego 2002; Perrone-Moisés 2004; Angotti-Salgueiro 2006).

Second, reflecting on the English language's spread after the Second World War, John Peter Cole, Brian Berry, and John Friedman have brought to the Brazilian Institute of Geography and Statistics the main novelties from the theoretical movement between 1967 and 1969. Two years later, Rio de Janeiro hosted the International Geography Union Meeting of the Commission on Quantitative Geography, and in 1974 David Harvey has visited a small but important group of new converts at Rio Claro University for the purpose of disseminating innovative methods (Camargo and Reis Júnior 2004).

Third, the latest geographical hegemony in last-century Brazil was exerted by radical geography, which has likewise experienced external stimuli. A dominant figure in this setting, Milton Santos was born in Brazil but obtained his PhD in Strasbourg in 1958, and he worked in Europe, Africa, North and South America before returning to his homeland in 1978 (see Ferretti and Viotto Pedrosa 2018). Referencing more texts in French and in English than in Portuguese as he did was unusual at that time. Yet it does not mean that translations were not relevant for Brazilian critical geographers. Originally published in 1976 and translated one year later into Portuguese by *Iniciativas Editoriais* in Portugal, Yves Lacoste's *La Géographie, ça sert, d'abord, pour faire la guerre* had a mythic role for them, for instance (Lacoste 1976). Due to the military regime (1964–1985), this book was only printed in Brazil in 1988.

For these reasons, when we try to trace how ideas travel, we cannot ignore the presence of languages and translations. They are so apparent that it seems normal to forget them, but this kind of amnesia becomes even more serious because both are geographic phenomena. Excluding some inspiring cases where circulation of knowledge and geographies of science have been highlighted (Ophir and Shapin 1991; Secord 2004; Livingstone 2003, 2005; Naylor 2005, 2005a), languages and translations have not been receiving the same attention, in scholarship on historical geography, in comparison with themes such as museums, laboratories, research networks, botanical and zoological gardens, scientific societies, and so on. Therefore, it is not a coincidence that many scholars claim that academy needs to pay more attention to translation (Schulte 1992; Rupke 2000; Venuti 2009). By contrast, from a Brazilian point of view, it is almost impossible to consider the history of geography without paying attention to languages and translations. This means that I do not adopt the perspective of the "national schools of geography," but the transnational one. I consider Brazilian geography as a sort of hybrid field where most of scholars have been trying to build their own pathways through imported inputs. In other words, Brazilian geographical tradition consists in thinking from, with, and beyond other intellectual backgrounds.

These circumstances help us to explain why translation is still important nowadays, for it represents a calling to a double utopia: a reaction from peripheries against centers and an egalitarian conversation between North and South. Now, translation is more relevant for South than for North; translating foreign texts is almost an obligation for us, a step that we cannot escape. Overall, it would be no exaggeration to state that "the history of Brazil is a history of translations and of linguistic change" (Barbosa and Wyler 2001 [1998]). As a result, a *geopolitics of translation* (see Spivak 2000 [1993]; Wright 2002; Cameron 2003) takes center stage, driving Brazilian journals to translate in the present and past as much as they can. On the one hand, a considerable proportion of these translations is a mere copy of foreign ideas, but on the other hand, the act of translating questions eurocentrism and its epistemological values in comparison with knowledges and languages from the South.

If translation (and language) is an ethical issue (Ricoeur 2004; Spivak 2000 [1999]), I need to interrogate its political consequences. Doing that requires a critical approach (Minca 2000; Garcia-Ramon 2003; Aalbers 2004; Desbiens and Ruddick 2006; Müller 2007; Germes and Husseini de Araújo 2016), for seemingly consensual terms current related to translation such as "diffusion of knowledge" or "cultural interexchange" masks epistemic conflicts and linguistic hegemonies. Essential for the circulation of science in peripheral countries, in mobilizing an array of authors, languages, theories, concepts, and methods, translations are more than random and neutral routines. Rather, they are social actions exposed to historical settings and ideological orientations. At the same time, bridge and fence, access and embargo, presence and absence, translations can be considered as a kind of migration. If I am aware that not everyone can travel the world according to your own preferences, I should deduce that translations are directly responsible for keeping asymmetries and hierarchies scientific on a global scale, although they may create open and democratic spaces of dialogue as well.

Quantitative geography in the IBGE journals: a study through its translations (1966–1982)

It is important to take into account that quantitative approaches fitted like a glove in a federal institute devoted so to geography and statistics as to territorial planning since the beginning. If we know that the 1940s and 1950s were consecrated to big expeditions across the countryside and the Northeast Region, as well as the Central Plateau targeting the building of Brasilia, it is not difficult to imagine how relevant fieldwork, landscape recognition, environmental surveys, and regionalization were put in practice by Francis Ruellan, Pierre Deffontaines, Pierre Monbeig, Leo Waibel, and Preston James. In the same vein, the degree of rationalization required by a mass society during the 1960s and 1970s was a response to demands such as urban growth prediction, energy and transport systems, regional inequalities, and economic and territorial integration. That is how mathematical models, spatial interactions, optimal location, networks, and computers become keywords used for modernizing geographical methods and for developing a Third World economy as well (see Faissol 1997). French names were replaced by English ones such as Brian J.L. Berry, John P. Cole, Paul R. Lohnes, David W. Harvey, and Chauncy D. Harris.

Even though geographies of reading, book, and print culture do not have many adepts in Brazil, there is no Brazilian geographical tradition without foreign impact. Nevertheless, these impacts are often released uncritically, and empirical data and sources are substituted by general assertions regarding a French heritage or an American influence. This is insufficient if I want to deepen how foreign geography has circulated in Brazil. After all whereas the *Revista Brasileira de Geografia* translated 90 papers in 1167 items in the years 1939–1996 (7.71%), the *Boletim Geográfico* reached the mark of 448 translations for a total of 2381 papers between 1943 and 1978 (18.81%). While the first journal has contained around 63 authors, 20 periodicals, and 29 translators distributed for the most part in English and French, but also in German and Latin, the second one has mobilized around 200 authors, 150 periodicals, and

60 translators distributed for the most part in French and English, but in Spanish, German, Italian, Romanian, and Japanese as well.

A kind of embassy in a pre-internet era, the language concentration does not exclude the transnational dimension of the *Boletim Geográfico* and one might presume that it represents one of the major 20th-century Western translation collections. Normally seen as a vehicle of scientific information for school teachers and the public in general, those numbers may also be useful to rethink about the secondary status attributed to the *Boletim Geográfico* in comparison with the *Revista Brasileira de Geografia*, which was consecrated to publish the main research results from the IBGE.

Boletim Geográfico's quantitative translations

Two years after the military coup that deposed the democratically elected president João Goulart in 1964, *Boletim Geográfico* has translated its first quantitative text. Under the responsibility of the IBGE geographer Lêda Pereira Ribeiro, Air Force Academy Harry W. Emrick's *A method for inter-regional comparisons (factor analysis)* (1966 [1965]) was originally published by *The professional geographer* a year earlier. However, it would take more three years until a new quantitative translation being published again: *Transportation and the growth of the São Paulo economy* (1969 [1968]), authored by Howard L. Gauthier in the *Journal of Regional Science*. In the same 1969, the last theoretical translation of the decade: *Wholesale-Retail Trade Ratios as Indices of Urban Centrality* [1961], by William R. Siddall in *Economic Geography*. There is no information about translators in both cases.

While quantitative geography emerged in the early 1950s, it took more than 10 years for the first translation into Portuguese. In general, the three translations mentioned earlier represent a shy start for a "revolution." Although in the course of the 1960s, the *Boletim Geográfico* has been translating geographers and economists engaging in applied geography such as Pierre George, Bernard Kayser, Michel Rochefort, and Jacques Boudeville – which might be seen as a kind of transition toward quantitative approaches (see Pedrosa 2017; Bonfim 2007) – the majority of the authors translated in the 1960s belong to the "regional school" like Pierre Deffontaines, Jean Tricart, André Cailleux, and Preston E. James. From 50 texts translated since the first quantitative one has appeared (1966–1969), only 3 were dedicated to spatial analysis (6%).

Things started to move in the following decade, and it is certain that the 1970s were the principal moment for that approach in Brazil. In 1971, the Department of Geography at the São Paulo State University (campus Rio Claro) created a journal especially dedicated to quantitative issues, the *Boletim de Geografia Teorética*. In the same year, in a partnership between IBGE and IGU, Rio de Janeiro city hosted the Quantitative Methods Commission Conference (as I have already pointed out), and consequently four articles were translated by the *Boletim Geográfico* from this event: *Transport planning and network analysis: a set of spatial models* (1971 [1971]) by Lalita Sen; *The Social Environment of Rio de Janeiro in 1960* by Fred Morris (1971 [1971]); *Measurement problems in geometric models, perception and preference* (1972 [1971]) by Donald Demko; and *Canonical correlation analysis in geography* (1972 [1971]) by D. Michel Ray and Paul Lohnes. Here again, it is impossible to know about translators.

Following the 1970s, *Boletim Geográfico* has continued to translate papers related to spatial analysis. It is possible to order them in two axes:

1 eight works on epistemological and bibliographical arguments as *Nouvelle frontière pour la recherche géographique* (Racine 1971 [1969]); *Prospects for the development of geographic sciences* (Leszczynski 1973 [1972]); *Quelques réflexions sur la recherche en géographie* (Bastié 1973 [1970]); *Models of economic development* (Keeble 1973 [1967]); *Regions, models and classes*[2] (Grigg 1973 [1967]); *The regionalizing ritual* (Fleming 1974 [1973]); *Un paradigme pour l'étude de l'organisation spatiale des sociétés* (Villeneuve 1974 [1972]); and *The use of the term "hypothesis" in geography* (Newman 1976 [1973]);
2 five case studies applying theory in practice as *Toward an Expanded Central-Place Model* (Thomas 1972 [1961]); *The market as an actor in the localization of industry in the United States* (Harris 1972 [1954]); *The use of multi-variable methods in regional geographical analysis* (Hautamäki 1974 [1970]); *Clustering of services in central places* (Bell, Lieber, and Rushton 1976 [1974]); and *Theoretical concepts and the analysis of agricultural land-use patterns in geography* (Harvey 1976 [1976]).

An analysis of two decades reveals that the strategy adopted was to prefer journals over book chapters and annals of congresses. There are 10 journals from five countries: Canada (*Cahiers de géographie du Québec*), France (*Acta Geographica*), Finland (*Fennia*), Scotland (*Scottish geographical magazine*) but mainly from the United States (*AAAG, Economic Geography, The professional geographer, The geographical review, Journal of Regional Science*), as well as the bilingual *IGU Bulletin/ Bulletin de l'UGI*. With five appearances, *AAAG* is the most translated journal. Related to language, English dominion is incontestable: from 20 texts, 17 were translated from English (85%) and just 3 were translated from French (15%). Concerning translators, four of six of them were IBGE geographers like the women Lêda Pereira Ribeiro and Edna Mascarenhas Sant'anna and the men Henrique Azevedo Sant'anna and Joaquim Quadros Franca, while the other two were professional translators as the men Patrice C.F.X. Wuillaume and Arnaldo Viriato de Medeiros. Unfortunately, there is no mention of translators in 11 papers, which occurs frequently in the *Boletim Geográfico*.

In the course of 1970–1978 (1978 was the latest year of the *Boletim Geográfico*, which was allegedly added to the *Revista Brasileira de Geografia* but this movement represents the decline of geography within the IBGE), from 85 translated texts 17 of them were devoted to spatial analysis (20%). From 1966 to 1976 (last year in which a quantitative text was translated), from 126 translated texts 20 of them corresponding to that approach (15.87%). To my mind, the first percentage is more meaningful than the second one because it displays how recurrent quantitative translations in the 1970s were. Conversely, the principal issue is that when we analyze the translations of that period as a whole we discover more than 50 authors from different methods, including Jean Tricart, Paul Claval, Carl Troll, Karlheinz Paffen, Harald Sioli, Erich H. Brown and even humanist geographers like Yi-Fu Tuan and Leonard Guelke, as well as themes such as Physical Geography, Environmental Matters, and Economic Development. Six papers were translated from the Commission on Regional Aspects of Development of the IGU, for example, whose approach is very close to the spatial analysis. Nevertheless, none of them have the same echo than quantitative matters. Additionally, in the 1970s, from 17 translations, 8 were published in the same year of the original version, 3 were published two years later, and 1 was published three years later, which means the haste in which the original versions were selected and spread among Brazilian readers.[3]

Revista Brasileira de Geografia's quantitative translations

In comparison to the *Boletim Geográfico*, the first quantitative translation published by the *Revista Brasileira de Geografia* would appear four years later in 1970: Brian J. L. Berry's and Gerald F. Pyle's *Major regions and types of agriculture in Brazil* (1970 [1969]). Published originally in a mimeographed version from the University of Chicago, its Brazilian version is part of an issue that especially focuses on quantitative geography. Composed of six papers, two of them are translations. The other one is John P. Cole, Speridião Faissol, and M.J. McCullagh's *The application of Markov chains to population projection in Brazil* (1970 [unknown]). There is no information about the translator in both cases.

Considered as a mark for spatial analysis in Brazil (Monteiro 2002), the editorial from that issue is an expressive synthesis of the challenges and ambitions in that times. Immersed in a military dictatorship under the aegis of President Médici and the Minister of Finance Delfim Netto, Brazilian government was promoting a brutal persecution against left-wing political groups while the "economic miracle" tried to persuade civil society that the military regime was much better than a communist spectrum that has been threatening Latin America since the Cuban Revolution in 1959. In close connection with the United States, IBGE geographers were politically and intellectually inspired by visitors such as Brian J.L. Berry, Howard Gauthier, and John P. Cole. By emphasizing topics such as objectivity, planning, and development, the "revolutionary" agenda had a double function: it did not disturb the regime and, in its own way, it contributed to explain social and economic problems. Indeed, the spatial analysis was attracting "great enthusiasm and interesting" and seemed to be a perfect tool to be applied to a country

> with almost a hundred million inhabitants and experiencing deep socio-economic changes. On the one hand, the process of development submitted to the tropical space frictions, which is characterized by uninhabited and overpopulated spaces, fast-growing metropolitan areas and zones affected by a macrocephaly responsible for deforming its natural expansion.

> On the other hand, modern and sophisticated techniques coming from other sciences and absorbing by geographers as weapons.
>
> (Revista Brasileira de Geografia, Editorial 1970, 32(4:4), my translation)

Although, the IBGE Publishing House translated from French to Portuguese books like *Visit of French Masters. Conferences and Lessons from Pierre George and Jean Tricart* in 1963 (no information about translators) and *Pierre George's Brazilian Conferences* in the 1970 (translated by IBGE geographers Olga Maria Buarque de Lima and Henrique Azevedo Sant' anna), there was no place for Marxist or Critical translations in the *Revista Brasileira de Geografia* between 1970 and 1982. In a similar way, Cultural and Humanistic approaches have received almost the same attention: the only text translated (no information about translators again) was Stephen S. Chang's *The role of cultural geographers in industrial decisions* (1978 [1975]). From 1970 to 1979, of 23 articles translated, 12 of them were devoted to quantitative geography (52.17%). From 1970 to 1982 (the last year in which a quantitative text was translated), for instance, of 31 translated texts, 16 of them correspond to that approach (51.61%). Just as the *Boletim Geográfico*, there is no unity among other translations published in the *Revista Brasileira de Geografia*, which means that it has deliberately concentrated its translations on quantitative subjects. They can be split into two categories:

1 ten articles on epistemological and bibliographical arguments as *A paradigm for modern geography* (Berry 1972); *Graphical Representation of a Matrix with Applications in Spatial Location* (Lindgren and Steinitz 1976 [1969]); *The logic of functional analysis* (Hempel 1976 [1959]); *Changes and Trends in Human Geography: Review. Reviewed Works: Human Geography: Evolution or Revolution? by Michael Chisholm; Patterns in Human Geography: An Introduction to Numerical Methods by David M. Smith* (Stanley 1978 [1977]); *Geography, geometry, and explanation* (Sack 1979 [1972]); *L'usage des comptabilités micro et macroéconomiques en géographie* (Derrieux-Cecconi 1979 [1975]); *Le teste de base de la répresentation graphique* (Bertin 1980 [1965]); *Ecology and Spatial Analysis* (Clarkson 1980 [1970]); *Applied Geography and Pragmatism* (Frazier 1981 [1978]); and *Principal components analysis and factor analysis in geographical research: some problems and issues* (Johnston 1982 [1977]);

2 six empirical cases (five of them emphasizing Brazil) such as *Major Regions and Types of Agriculture in Brazil* (Berry and Pyle 1970 [1969]); *The application of Markov chains to population projection in Brazil* (Cole, Faissol, and McCullagh 1970 [unknown]); *An analysis of inequalities income growth in Brazil according to information theory concepts* (Semple and Gauthier 1971 [1971]); *The Social Environment of Rio de Janeiro in 1960* (Morris 1973 [1970]); *Structural Models of Retail Distribution: Analogies with Settlement and Urban Land-Use Theories* (Davies 1976 [1972]); and *Rank size distribution, city size hierarchies and The Beckmann model, some empirical results* (Suárez-Villa 1979 [1979]).

Concerning sources, languages, and translators, we note six different journals from four different countries: *AAAG* and *The Geographical Survey* (United States); *Transactions. Institute of British Geographers* and *The Geographical Journal* (United Kingdom); *Annales de Géographie* (France); and *South African Geographical Journal* (South Africa). There are also three book chapters and three papers, as well as a master dissertation and an article translated from the already reported "Quantitative Methods Commission Conference" from IGU in 1971. Taken together, from 16 translated texts, 14 of them were in English (87.5%) and just two of them were in French (12.5%). This evidence accounts for the Anglo-Saxon hegemony in terms of quantitative geography performed by *Revista Brasileira de Geografia*.

Most of the time, though, it is impossible to know who was responsible for all this work. Unhappily, just five of them have indications about translators. They are Patrice Charles F. X. Wuillaume, professional translator who had worked for *Boletim Geográfico* in the 1970s; Joaquim Quadros Franca, IBGE revisor and one of the main translators in this latter journal; Ângela Maria Rocha Lima Diego, official linked to the IBGE Publishing Center; Carlos Ernesto da Silva Lindgren, engineer and professor at Rio de Janeiro Federal University; and Odeibler Santo Guidugli, PhD in geography at that time and university professor at São Paulo State University as a result. It is interesting to observe that Lindgren is both an author and a translator of your own article written with Carl Steinitz, and I presume that Speridião Faissol – an IBGE geographer and the greatest name of the quantitative movement in Brazil – has made the same in his article written with Cole and McCullagh, but it is difficult to be sure of this.

In scrutinizing these materials, a lack of information with reference to original sources and translators represents a bigger problem and I suppose that this kind of research would be almost impossible if it were not for all files from different journals from many parts of the world available on the Internet.

Nevertheless, one of them is particularly unusual because it discloses how translation may shed new light on the circulation of science. Published in the *Revista Brasileira de Geografia* in 1972, the only information concerning Brian Berry's *A paradigm for modern geography* refers to the translator: Patrice Wuillaume. Thanks to the fact that we are facing a well-known work, it was easy to find the source: *Directions in geography*, edited by Richard Chorley and printed by Methuen in 1973 (Chorley 1973). The only problem is as follows: How a translation could be published *before* the original version? According to Brian Berry,[4] the translation into Portuguese was faster than the publication of the book chapter in London. The enigma was rapidly solved but not the process behind it. This episode displays two things at least: first, the rush to spread that material on the part of the Brazilian geographers; second, the extent of some languages and vehicles in comparison to others. Notwithstanding the Brazilian version had been published earlier than the English one, both it and the *Revista Brasileira de Geografia* are absolutely unknown within the canons of the quantitative geography.

At the end of the 1970s, *Revista Brasileira de Geografia* is still a conservative periodical. From the other side, organized by the *Associação dos Geógrafos Brasileiros* [*Association of Brazilian Geographers*] in 1978 the National Conference at Fortaleza is considered a key moment where Marxist geographers start to contest IBGE geographers and, led by Milton Santos, will "take the power" during the next 20 years. Nonetheless, although *Revista Brasileira de Geografia* has dedicated not a word to that Conference and has continued to translate quantitative texts until 1982, at the beginning of the 1980s, we identify the presence of social topics such as education and inequalities (Strauch 1980), income and urban poverty (Geiger 1980), residential segregation (Vetter 1981), housing conditions of low-income populations (Lima 1981), urban education level (Davidovich and Cardoso 1982), supplementary education, labor market and urban poverty (Cardoso 1982, 1982a), immigration and slums (Bezerra and Cruz 1982), the use of the social division of labor concept in order to understand the Brazilian urban system (Fredrich and Davidovich 1982), capitalism and small-scale agriculture (Brito and Silva 1982), and even environmental degradation provoked by the modernization of agriculture (Romeiro and Abrantes 1982).

What is also interesting here is that some of these authors, like Geiger, Fredrich, and Davidovich, have been some of the main enthusiasts of quantitative methods in the IBGE. In this sense, Guidugli represents a singular case for, if on one hand, he translated John W. Frazier's *Applied Geography and Pragmatism* in 1981 (1978), on the other hand, he wrote in the following issue a friendly review on Derek Gregory's *Ideology, Science and Human Geography* (1981 [1979]) where we can read that "By questioning many positivist assertions, first of all the author sought to reintroduce man into the studies of geography in your due place. Such a task was done with a certain leg" (Santo Guidugli 1981, 453, my translation). The end of an era had arrived.

Conclusions

As a Brazilian geographer, I have been studying and writing within an atmosphere where translations have been playing a significant function. This is still an underestimated subject among my pairs. However, after radical criticism in the 1980s, works like those of Reis Jr. (2007) and Lamego (2010) help to realize the development of spatial analysis in Brazil in broader terms. Thus I have decided to study the subjects in exploring two of the main Brazilian periodicals in the 20th century: *Boletim Geográfico* and *Revista Brasileira de Geografia* – both edited by the Brazilian Institute of Geography and Statistics, one of the two most important quantitative centers in Brazil along with the São Paulo State University (campus Rio Claro).

In taking together the two periodicals, I noted a period of 16 years from the first quantitative translations in 1966 to the last one in 1982. In the meantime, they mobilized 36 texts, 42 authors, and 15 journals from seven countries and nine translators. There is an equilibrium between epistemological topics and case studies, which is consistent with the need of theory for empirical inquiries advocated by the quantitative revolution. A similar equilibrium can be found concerning the gender of translator: five men and four women, which is also consistent with a strong presence of women among the translators of that periodicals. By valuing journal papers more than book chapters and conferences, I confirmed its capacity in disseminating information quickly and efficiently. Nor should we ignore the high number of translated articles without translators. If it is implausible that a scientific paper might be published without a signature, what makes us think that an anonymous translation is a normal procedure?

Inspired by a thought-provoking literature from the translation studies (Venuti 1995; Rupke 2000; Simon 2009; Bachmann-Medick 2009; Italiano 2016), whose teachings are correlated with ideas of a particular geography of translation and its inherently political facet, the answer is as follows: for the most part of geographers, translation is still seen as a mere transcription, a copy. This is an oversimplification. Rather, translation is a way by which we can rule know-how, practices, and concepts. It means the power of choosing what, who, and which themes will be spread for a wide public and supposedly beyond linguistic boundaries. The reasons could be pedagogical, intellectual, or political, but the fact is that some translators, institutions, and journals act as social agents able to impose their wills upon the scientific field as a whole. It is not a mechanical movement, *bien entendu*, but what is at stake here is the strong involvement of the IBGE journals with the purpose of rendering Anglo-Saxon quantitative geographers into Portuguese. For a total of 36 quantitative translated texts in both journals, 31 were rendered from English (86.22%) and just 5 of them (13.88%) were translated from French. The main source is the *AAAG*.

So if I ignore translations, I contribute to naturalize the circulation of science and its hierarchies around the world. As a result, I do not understand the geopolitical outcomes of language on a peripheral country as Brazil, where the French culture reigned from the 19th century to 1945 and the American way of life pervades intellectual sphere since the end of the Second World War. The positive reception of the quantitative geography is a question of scientific innovation (the French regional method, which dominated Brazilian geography from the 1930s to the 1960s, was considered out of date), but also an issue of politics and economy in a country aligned with the United States during the Cold War and compromised with a military dictatorship from 1964 to 1985. As one of a big federal institution for planning, spatial analysis represented more than a scientific wave for the IBGE. It is also an opportunity to participate in the capitalist development. Computers, mathematical models, and projections composed an arsenal from which geographers could test this new background in order to supply maps and databases for public politics and pedagogical text material for school and university teachers. In both circumstances, translating was decisive.

Acknowledgments

I would like to express my gratitude to all those present in the Workshop on Histories of Quantitative Revolutions in Geography in Kiel on September 23–25, 2019, mostly Ferenc Gyuris, Boris Michel, and Katharina Paulus. Brian Berry helped me in a way that any archive could do so, and Federico Ferretti and Paul Claval not only rectified my English but also gave me good suggestions.

Annexes

Boletim Geográfico's quantitative translations (1966–1976)

YEAR	*AUTHOR*	*TITLE*	*SOURCE*	*TRANSLATOR*
1966	Harry W. Emrick	A method for inter-regional comparisons (factor analysis)	The Professional Geographer 1965	Lêda Pereira Ribeiro
1969	Howard L. Gauthier	Transportation and the growth of the São Paulo Economy	Journal of Regional Science 1968	Unknown
1969	William R. Siddall	Wholesale-Retail Trade Ratios as Indices of Urban Centrality	Economic Geography 1961	Unknown
1971	Jean-Bernard Racine	Nouvelle frontière pour la recherche géographique	Cahiers de géographie du Québec 1969	Unknown
1971	Lalita Sen	Transport planning and network analysis: a set of spatial models	Quantitative Methods Commission, IGU, Brazil, 1971	Unknown
1971	Fred B. Morris	The Social Environment of Rio de Janeiro in 1960	Quantitative Methods Commission, IGU, Brazil, 1971	Unknown
1972	Donald Demko	Perception and preference structures with respect to spatial choices/Problemas de mensuração em modelos geométricos, da percepção e da preferência	Quantitative Methods Commission, IGU, Brazil, 1971	Unknown
1972	D. Michel Ray and Paul Lohnes	Canonical correlation analysis in geography	Quantitative Methods Commission, IGU, Brazil, 1971	Unknown
1972	Edwin N. Thomas	Toward an Expanded Central-Place Model	The Geographical Review 1961	Unknown

(*Continued*)

(Continued)

YEAR	*AUTHOR*	*TITLE*	*SOURCE*	*TRANSLATOR*
1972	Chauncy D. Harris	The market as a factor in the localization of industry in the United States	AAAG 1954	Patrice C.F.X. Wuillaume
1973	Stanislaw Leszczynski	Prospects for the development of geographic sciences	IGU Bulletin 1972	Joaquim Quadros Franca
1973	D.E. Keeble	Models of economic development	Models in Geography ed. by Chorley & Haggett 1967	Arnaldo Viriato de Medeiros
1973	David Grigg	Regions, models and classes	Models in Geography ed. by Chorley & Haggett 1967	Arnaldo Viriato de Medeiros
1973	Jean Bastié	Quelques réflexions sur la recherche en géographie	Acta Geographica 1970	Edna Mascarenhas Sant'anna
1974	Douglas K. Fleming	The regionalizing ritual	Scottish Geographical Magazine 1973	Unknown
1974	Lauri Hautamäki	The use of multi-variable methods in regional geographical analysis	Fennia 1970	Edna Mascarenhas Sant'anna
1974	Paul Y. Villeneuve	Un paradigme pour l'étude de l'organisation spatiale des sociétés	Cahiers de géographie du Québec 1972	Henrique Azevedo Sant'anna
1976	Thomas L. Bell, Stanley R. Lieber, and Gerard Rushton	Clustering of services in central places	AAAG 1974	Unknown
1976	David Harvey	Theoretical concepts and the analysis of agricultural land-use patterns in geography	AAAG 1976	Joaquim Quadros Franca
1976	James L. Newman	The use of the term "hypothesis" in geography	AAAG 1973	Unknown

Source: Guilherme Ribeiro

Revista Brasileira de Geografia quantitative translations (1970–1982)

YEAR/ ISSUE	*AUTHOR*	*TITLE*	*SOURCE*	*TRANSLATOR*
1970 n.4	Brian J. L Berry Gerald F. Pyle	Major Regions and Types of Agriculture in Brazil	The University of Chicago 1969, mimeographed	Unknown
1970 n.4	John P. Cole Speridião Faissol M.J. McCullagh	The application of Markov chains to population projection in Brazil	Unknown	Unknown
1971 n.4	R. K. Semple H. L. Gauthier	An analysis of inequalities income growth in Brazil according to information theory concepts	Quantitative Methods Commission, IGU, Brazil, 1971	Unknown
1972 n.3	Brian Berry	A paradigm for modern geography	Directions in Geography ed. by Chorley 1973	Patrice Charles F. X. Wuillaume
1973 n.1	Fred B. Morris	The Social Environment of Rio de Janeiro in 1960	Chicago University, May 1970	Unknown
1976 n.1	C. Ernesto S. Lindgren Carl Steinitz	Graphical Representation of a Matrix with Applications in Spatial Location	Harvard Papers in Theoretical Geography, "Geography and the Properties of Surfaces" Series, Paper Number Thirty-Three, 1969	C. Ernesto S. Lindgren
1976 n.2	Carl G. Hempel	The logic of functional analysis	Symposium on Sociological Theory (book) 1959	Joaquim Quadros Franca
1976 n. 4	R. L. Davies	Structural Models of Retail Distribution: Analogies with Settlement and Urban Land-Use Theories	Transactions of the Institute of British Geographers 1972	Unknown
1978 n.2	M. J. Stanley	Changes and Trends in Human Geography: Review Reviewed Works: Human Geography: Evolution or Revolution? by Michael Chisholm; Patterns in Human Geography: An Introduction to Numerical Methods by David M. Smith	The Geographical Journal 1977	Unknown

(*Continued*)

(Continued)

YEAR/ ISSUE	*AUTHOR*	*TITLE*	*SOURCE*	*TRANSLATOR*
1979 n. 1–2	Robert David Sack	Geography, geometry, and explanation	AAAG 1972	Unknown
1979 n. 4	Régis Derrieux-Cecconi	L'usage des comptabilités micro et macroéconomiques en géographie	Annales de Géographie 1975	Unknown
1979 n. 4	Luis Suárez-Villa	Rank size distribution, city size hierarchies and The Beckmann model, some empirical results	Ithaca, NY: Department of City and Regional Planning in conjunction with the Program in Urban and Regional Studies, Cornell University, 1979	Unknown
1980 n.1	Jacques Bertin	Le teste de base de la réprésentation graphique	Sémiologie Graphique (livre). 2ème édition. (1973 [1965]).	Unknown
1980 n. 2	James D. Clarkson	Ecology and Spatial Analysis	AAAG 1970	Unknown
1981 n.2	John W. Frazier	Applied Geography and Pragmatism	The Geographical Survey 1978	Odeibler Santo Guidugli
1982 n. 4	R. J. Johnston	Principal components analysis and factor analysis in geographical research: some problems and issues	South African Geographical Journal 1977	Ângela Maria Rocha Lima Diego (CEDIT)

Source: Guilherme Ribeiro

Notes

1 Although translations do not have much relevance within this literature review, who could deny how important the translation from Swedish to English of Hägerstrand's *Innovation Diffusion as a Spatial Process* (1967 [1953]) by Alan Pred was? (see Pred 1984, 98).

2 The book in which this article was published – *Models in geography* (Chorley and Haggett 1967) – was translated into Portuguese in 1974 and helped to diffuse the "revolution." Arnaldo Viriato de Medeiros was the professional translator and the book was published in a partnership between the publishing houses Ao Livro Técnico and Editora da Universidade de São Paulo.

3 See the case of Brian Berry in the next pages.
4 As per his email reply on September 26, 2019.

References

Aalbers, M. B. (2004): Creative Destruction through the Anglo-American Hegemony: A Non-Anglo-American View on Publications, Referees and Language. In *Area*, 36(3), 319–322.

Angotti-Salgueiro, H. (org.) (2006): *Pierre Monbeig e a Geografia Humana Brasileira: A Dinâmica da Transformação*. Bauru: Edusc.

Anonymous (1970): "Editorial". In *Revista Brasileira de Geografia*, 32(4), 3–4.

Bachmann-Medick, D. (2009): The Translational Turn. In *Translation Studies*, 2(1), 2–16.

Bachmann-Medick, D. (2016 [2015]): *Cultural Turns: New Orientations in the Study of Culture*. Berlin/Boston: De Gruyter, translated by Adam Blauhut.

Barbosa, H. G.; Wyler, L. (2001 [1998]): Brazilian Tradition. In Baker, M. (Ed.): *Routledge Encyclopedia of Translation Studies*. London and New York: Routledge, 338–344.

Barnes, T. J. (2004): Placing Ideas: Genius Loci, Heterotopia and Geography's Quantitative Revolution. In *Progress in Human Geography*, 28(5), 565–595.

Barnes, T. J. (2016): The Odd Couple: Richard Hartshorne and William Bunge. In *The Canadian Geographer/Le Géographe Canadien*, 60(4), 458–465.

Bezerra, V. M. d'A. C.; Cruz, J. M. (1982): Imigração e Favelas: o Caso do Rio de Janeiro em 1970. In *Revista Brasileira de Geografia*, 44(2), 357–367.

Billinge, M.; Gregory, D.; Martin, R. (Eds.) (1984): *Recollections of a Revolution: Geography as a Spatial Science*. New York: The Macmillan Press, 86–103.

Bonfim, P. R. de A. (2007): *A Ostentação Estatística (um Projeto Para o Território Nacional: Estado e Planejamento no Período Pós-64)*. São Paulo: Tese de Doutorado, Geografia Humana, FFLCH/USP.

Bourdieu, P. (1976): Le Champ Scientifique. In *Actes de la Recherche en Sciences Sociales*, 2(2–3), 88–104.

Brito, M. do S.; Silva, S. T. (1982): O Papel da Pequena Produção na Agricultura Brasileira. In *Revista Brasileira de Geografia*, 44(2), 191–261.

Bunge, W. (1979): Perspective on *Theoretical Geography*. In *AAAG*, 69(1), March, 169–174.

Camargo, J. C. G.; Reis, D. F. da C., Jr. (2004): Considerações a Respeito da Geografia Neopositivista no Brasil. In *Geografia* (Rio Claro), 29, 355–382.

Cameron, D. (2003): Foreign Exchanges: The Politics of Translation. In *Critical Quaterly*, 45(1–2), 215–219.

Cardoso, M. F. T. (1982): Características Sócio-Espaciais de Uma Clientela do Ensino Supletivo. In *Revista Brasileira de Geografia*, 44(1), 163–179.

Cardoso, M. F. T. (1982a): A Inserção Precoce no Mercado de Trabalho e a Clientela do Ensino Supletivo: um Estudo da Pobreza Urbana. In *Revista Brasileira de Geografia*, 44(2), 331–355.

Charlton, M. (2008): Location Analysis in Human Geography (1965): Peter Haggett. In Hubbard, P.; Kitchin, R.; Valentine, G. (Eds.): *Key Texts in Human Geography*. London: SAGE, 17–24.

Chorley, R. J. (Ed.) (1973): *Directions in Geography*. London: Methuen.

Chorley, R. J.; Haggett, P. (Ed.). (1967): *Models in Geography*. London: Methuen.

Cronin, M. (2003): *Translation and Globalization*. London and New York: Routledge.

Cronin, M. (2010): Knowing One's Place: Travel, Difference and Translation. In *Translation Studies*, 3(3), 334–348.

Davidovich, F.; Cardoso, M. F. T. C. (1982): Resultados Preliminares de um Estudo Geográfico Sobre Aglomerações Urbanas no Brasil: Análise do Nível de Instrução. In *Revista Brasileira de Geografia*, 44(1), 89–135.

Delfosse, C. (1998): Le Rôle des Institutions Culturelles et Des Missions à l'étranger Dans la Circulation des Idées Géographiques. L'exemple de la Carrière de Pierre Deffontaines (1894–1978). In *Finisterra*, 33(65), 147–158.

Desbiens, C.; Ruddick, S. (2006): Speaking of Geography: Language, Power, and the Spaces of Anglo-Saxon "Hegemony". In *Society and Space*, 24, 1–8.

Faissol, S. (1997): Cinquenta Anos de Geografia. Entrevista Feita por Helion Póvoa Neto e João Rua. In *GeoUERJ*, 1, 55–70.

Ferretti, F.; Viotto Pedrosa, B. (2018): Inventing Critical Development: A Brazilian Geographer and His Northern Networks. In *Transactions of the Institute of British Geographers*, 43(4), 703–717.

Foucault, M. (1971): *L'ordre du Discours*. Paris: Gallimard.

Fredrich, O. M. B. de L.; Davidovich, F. (1982): A Configuração Espacial do Sistema Urbano Brasileiro Como Expressão no Território da Divisão Social do Trabalho. In *Revista Brasileira de Geografia*, 44(4), 541–590.

Garcia-Ramon, M.-D. (2003): Globalization and International Geography: The Questions of Languages and Scholarly Traditions. In *Progress in Human Geography*, 27(1), 1–5.

Geiger, P. P. (1980): Fluxos Interestaduais de Vazamento de Renda e Pobreza Urbana. In *Revista Brasileira de Geografia*, 42(3), 477–516.

Germes, M.; Husseini de Araújo, S. (2016): For a Critical Practice of Translation in Geography. In *ACME: An International Journal for Critical Geographies*, 15(1), 1–14.

Goodchild, M. F. (2008): Theoretical Geography (1962): William Bunge. In Hubbard, P.; Kitchin, R.; Valentine, G. (Eds.): *Key Texts in Human Geography*. London: SAGE, 9–16.

Gyuris, F. (2018): Problem or Solution? Academic Internationalization in Contemporary Human Geography in East Central Europe. In *Geographische Zeitschrift*, 106(1), 38–49.

Italiano, F. (2012): Translating Geographies: The Navigatio Sancti Brendani and Its Venetian Translation. In *Translation Studies*, 5(1), 1–16.

Italiano, F. (2016): *Translation and Geography*. London and New York: Routledge.

Johnston, R. (2008): Explanation in Geography (1969): David Harvey. In Hubbard, P.; Kitchin, R.; Valentine, G. (Eds.): *Key Texts in Human Geography*. London: SAGE, 25–32.

Keighren, I. M.; Abrahamsson, C.; Della Dora, V. (2012): On Canonical Geographies. In *Dialogues in Human Geography*, 2(3), 296–312.

Lacoste, Y. (1976): *La Géographie, ça Sert d'abord, à Faire la Guerre*. Paris: Maspéro.

Lamego, M. (2010): *Práticas e Representações da Geografia Quantitativa no Brasil: a Formação de uma Caricatura*. Rio de Janeiro: Tese de Doutorado, PPGG/UFRJ.

Latour, B. (1987): *Science in Action: How to Follow Scientists and Engineers through Society*. Cambridge, MA: Harvard University Press.

Lima, M. H. B. de (1981): Condições de Habitação da População de Baixa Renda da Região Metropolitana do Rio de Janeiro. In *Revista Brasileira de Geografia*, 43(4), 605–630.

Livingstone, D. (2003): *Putting Science in Its Place*. Chicago: University of Chicago Press.

Livingstone, D. (2005): Science, Text and Space: Thoughts on the Geography of Reading. In *Transactions of the Institute of British Geographers*, 30, 391–401.

Machado, M. S. (2009): *A Construção da Geografia Universitária no Rio de Janeiro*. Rio de Janeiro: Apicuri.

Martin, G. J. (1972): *All Possible Worlds: A History of Geographical Ideas*. New York/Oxford: Oxford University Press.

Mbembe, A. (2013): *Critique de la Raison Noire*. Paris: La Découverte.
Miceli, S. (org.). (1989): *História das Ciências Sociais no Brasil*. São Paulo: Vértice, Editora Revista dos Tribunais: IDESP.
Minca, C. (2000): Venetian Geographical Praxis. In *Society and Space*, 18, 285–289.
Monteiro, C. A. F. (2002): *A Geografia no Brasil ao Longo do Século XX: um Panorama*. São Paulo: AGB, Borrador.
Moraes, J. G. V. De; Rego, J. M. (2002): *Conversas com Historiadores Brasileiros*. São Paulo: Editora 34.
Müller, M. (2007): What's in a Word? Problematizing Translation between Languages. In *Area*, 39(2).
Naylor, S. (2005): Historical Geography: Knowledge, in Place and on the Move. In *Progress in Human Geography*, 29, 626–634.
Naylor, S. (2005a): Introduction: Historical Geographies of Science: Places, Contexts, Cartographies. In *British Society for the History of Science*, 38(1), 1–12.
Novaes, A. R. (2015): Celebrations and Challenges: The International at the 16th International Conference of Historical Geographers, London, July 2015. In *Journal of Historical Geography*, 50, 106–108.
Ophir, A.; Shapin, S. (1991): The Place of Knowledge: A Methodological Survey. In *Science in Context*, 4(1), 3–21.
Paris, E. (1999): *La Genèse Intellectuelle de L'oeuvre de Fernand Braudel*. Athènes: Institute de Recherches Néohelléniques/FNRS.
Pedrosa, B. V. (2017): A Recepção da Teoria dos Polos de Crescimento no Brasil. In *Terra Brasilis* (Nova Série), 9, 1–15.
Perrone-Moisés, L. (org.). (2004): *Do Positivismo à Desconstrução: Idéias Francesas na América*. São Paulo: Edusp.
Pred, A. (1984): From Here and Now to There and Then: Some Notes on Diffusions, Defusions and Disillusions. In Billinge, M.; Gregory, D.; Martin, R. (Eds.): *Recollections of a Revolution: Geography as a Spatial Science*. New York: The Macmillan Press, 86–103.
Quijano, A. (2000): Coloniality of Power, Eurocentrism, and Latin America. In *Nepantla: Views from South*, 1(3), 533–580.
Reis, D. F. da C., Jr. (2007): Cinqüenta Chaves. In *O Físico Pelo viés Sistêmico, o Humano nas Mesmas Vestes . . . e uma Ilustração Doméstica: o Molde (Neo) Positivista Examinado em Textos de Antonio Christofoletti*. Campinas, SP: Tese (Doutorado) Universidade Estadual de Campinas, Instituto de Geociências.
Ricoeur, P. (2004): *Sur la Traduction*. Paris: Bayard.
Romeiro, A. R.; Abrantes, F. J. (1982): Degradação Ambiental e Ineficiência Energética (o Círculo Vicioso da "Modernização" Agrícola). In *Revista Brasileira de Geografia*, 44(3), 477–495.
Rupke, N. (2000): Translation Studies in the History of Science: The Example of Vestiges. In *British Journal for the History of Science*, 33(2), 209–222.
Santo Guidugli, O. (1981): Geografia Humana: Ciência ou Ideologia? (Comentário Bibliográfico). In *Revista Brasileira de Geografia*, 43(3), 451–453.
Schulte, R. (1992): Translation and the Academic World. In *Translation Review*, 38–39(1), 1–2.
Secord, J. (2004): Knowledge in Transit. In *Isis*, 95(4), 654–672.
Simon, S. (2009): Response. In *Translation Studies*, 2(2), 208–213.
Spivak, G. C. (2000 [1993]): The Politics of Translation. In Venuti, L. (Ed.): *The Translation Studies Reader*. London and New York: Routledge, 397–416.
Spivak, G. C. (2000 [1999]): Translation as Culture. In *Parallax*, 6(1), 13–24.

Strauch, L. M. de M. (1980): Educação e Comportamento Espacial. In *Revista Brasileira de Geografia*, 42(1), 31–51.

Venuti, L. (2009): Translation, Intertextuality, Interpretation. In *Romance Studies*, 27(3), 157–173.

Venuti, L. (1995): *The Translator's Invisibility: A History of Translation*. London: Routledge.

Vetter, D. M. (1981): A Segregação Residencial da População Econômicamente Ativa na Região Metropolitana do Rio de Janeiro, Segundo Grupos de Rendimento Mensal. In *Revista Brasileira de Geografia*, 43(4), 587–604.

Wright, M. W. (2002): The Scalar Politics of Translation. In *Geoforum*, 33, 413–414.

5 Digitality

Origins, or the stories we tell ourselves[1]

Matthew W. Wilson

It's very late, but just perhaps not too late to be decent.

– William Warntz, "Preface"

This disassociation between an art and its history is always ruinous.

– Gilles Deleuze, "Two Regimes of Madness"

The "Harvard Papers in Theoretical Geography" were edited by Bill Warntz (1922–88), the second director of the Harvard Laboratory for Computer Graphics and Spatial Analysis (LCGSA). He opens the papers in 1968 with an introduction to Bill Bunge's (1928–2013) manuscript on spatial analysis and the trials of Fred Schaefer (1904–53), a tribute to the deceased geographer from the University of Iowa (Bunge 1968). Schaefer advocated a scientific approach to geography, and his "Exceptionalism in Geography," published posthumously in 1953, was a momentous attempt to critique an unchallenged prioritization of the region in spatial inquiry (Schaefer 1953).[2] Bunge's manuscript on Schaefer has the feel of an obituary with a biting, impassioned edge. Warntz's introduction describes the difficulty with which Bunge was able to publish his research, paralleling the difficulties Schaefer experienced in publishing his "Exceptionalism," a couple of decades earlier. To begin the "Harvard Papers" with controversy underlines the affective moment of 1960s American geography. Areas of specialty were being reconfigured. Departments were being reorganized with the emergence of new computing power. So perhaps it is also very late to tell the story of another key figure from this time period, often lost in the shuffle: Howard Fisher. His efforts, some intentional and others happenstance, condition the beginnings of the digital map as we interact with it today.[3] If the tracing of maps is a key part of learning the craft of hand-drawn cartography, how might we trace the digital map?

In 1938, Erwin Raisz stated what would become obvious, if opaque, "Maps constitute an important part of the equipment of modern civilization" (Raisz 1938). By the early 1960s, Raisz would be in attendance at luncheons at Harvard's Faculty Club organized by Howard Fisher on the topic of computer mapmaking. Indeed, our well-told origin stories associated with GIS focus

DOI: 10.4324/9781003122104-5

there and then, around experiments in computer-supported cartography at the Laboratory for Computer Graphics (and Spatial Analysis) founded by Fisher in 1965. Stories of the history of the digital map emphasize the lab and, in particular, the productivity of men at the LCGSA.[4]

In what follows, I lean on the Fisher papers to understand the provenance of the LCGSA, the specific problems that were meant to be solved, and the productivity of experimentation in cartography, while proposing lines of inquiry that enlarge what digitality would come to mean. The point is to thicken the digital lines drawn with computation, recognizing these moments as technosocial – not in order to constitute a rigid intellectual history but to disrupt easy origin stories with the cul-de-sacs of experimentation and failure, tenuous allies and adversaries, and the fragility of thought and action.

Problem solving

Fisher was born in 1903 and died in early 1979. After earning his Bachelor of Science degree at Harvard in 1926, and spending just a year studying architecture, he moved to Chicago to found a company called General Houses Inc. He set about creating solutions to the problem of housing, adapting factory methods to the development of prefabricated homes, concrete curtain wall systems, and factory-made and site-assembled stair systems. Unfortunately, the Great Depression would mean that General Houses would build and sell few prefabricated houses. Fisher instead found a variety of work consulting on shopping centers and freeway bypasses as a new automobility swept municipal planning and development.

His collision with the use of the computer for mapmaking would happen at Northwestern University, where he had been a lecturer since 1957. And so the story goes that in 1964 Fisher enlisted the support of Betty Benson (1924–2008), a computer programmer, to develop the Synagraphic Mapping System (SYMAP).[5] This work captured the attention of those at the Harvard GSD and Dean Josep Lluís Sert. At Harvard, Fisher would question some of the assumptions of midcentury cartography. He attempted to rigorously understand the impact of the computer on thematic cartography, and with these new innovations, he experimented with the various conventions of subject selection and generalization.

His approach ruffled feathers, as can be seen in this transcribed conference session from 1970 between Fisher, George Jenks, of the University of Kansas, and Arthur Robinson, of the University of Wisconsin:

Professor Fisher:	You are the leading expert on this, and we don't know as much as you. As a result of your comments, we will go back and run more [computations] and see how we like it.
Professor Jenks:	Good.
Chairman Robinson:	I think we have taken care of that subject.[6]

While in only an excerpt, we can sense the tension between the cartographic establishment represented by Jenks and Robinson and the new computer cartography. Fisher defers to Jenks on the topic of generalization. Jenks's curt reply is followed by Robinson, who hopes to move the conversation along. In these moments, Fisher's first reaction is not to stubbornly dig in but to lean on a relentless empiricism and experimentation. Contra Robinson, the subject of map generalization and the persistence of data were far from handled.

University subjects

Return to 1930, the year Fisher founded General Houses Inc. in Chicago. In that year at Harvard, Alexander Hamilton Rice (1875–1956) entered into an agreement with President Lowell, for the establishment of a "School of Geography." The agreement established the intended focus of the school: "the teaching of fundamentals of geographical science and their application to pure regional geography, particularizing, especially, in geographic and physiographic problems and their relation with complementary ontographic problems." The school would be built and operated through funds from Rice, and his gifts would continue as long as Rice was able and interested, and that, "upon his death or retirement" (of importance to this story), such funds would ensure the continuation of geographical work within the school.[7]

It is generally well known that much of Rice's fortune was from Eleanor Widener, who is weaved into stories told on Harvard Yard connecting the monstrosity of the Titanic with that of Widener Memorial Library, which opened in 1915, to memorialize Widener's son and husband, who perished with the sinking of the British passenger liner. Indeed, it was reported at the dedication ceremony of the building on June 24 that Rice met Widener. Widener's fortune would support Rice's interest in the building of the Institute of Geographical Exploration on Divinity Avenue. Pure regional geography would have a home, and the chief cartographer would be Erwin Raisz, who was introduced to Rice by William Morris Davis.[8]

Raisz is credited with the first general cartography textbook of the modern era in English, published in 1938.[9] His students were trained in pen and ink cartography, and his drawings – including his maps – highlight a hand-born craft in its final moments of the 20th century. Raisz was committed to a democratization of geographic representation, with maps and other graphics that could be read at the surface without a belabored expertise.[10] Raisz and Hamilton Rice's twilight years would see the demise of the institute and the scattering of geographers from Harvard, in a story that borders on lore. Neil Smith relates the tale of two university presidents, Isaiah Bowman, geographer and then president of Johns Hopkins, and president of Harvard Jim Conant (Smith 1987; see also DeVivo 2014). But before Conant would speak those words in 1948, that geography was not a university discipline, there would be discussions, Smith argues, of homophobia and an "unsavory" relationship between the chair of the department, Derwent Whittlesey, and his friend and lecturer in the department, Harold Kemp.

Geography at Harvard came to an abrupt end in 1948.[11] Furious, Rice pulled his endowment arrangements, just eight years before his death in 1956 would have perhaps made the gift permanent. In the same year, Whittlesey's death would leave Harvard without a geography professor and coursework.[12] Indeed, from 1956 until 1964, one might argue that little if any geographical work was explicitly under way at Harvard.

Harvard's loss would be quickly overshadowed by revolutionary stirrings at the University of Washington, where a new cohort of geography graduate students arrived to study with Edward Ullman, a former geography professor at Harvard who had left in 1951. Geographers like Dick Morrill, Duane Marble, and John Nystuen were drawn to Ullman.[13] However, as Ullman was away on fellowship, many took up study with Bill Garrison – who introduced them to statistical methods in mapping. This rubbed against an established regional geography in the discipline, igniting what is now fabled as the quantitative revolution. Geographers from the University of Washington, including Waldo Tobler and Bill Bunge, added fuel to the fire (see Barnes 2001; Morrill 1984).

For these "Garrison Raiders," geography was "a basic science." However, Bunge found that geography's claim to be a science would require more explicit treatment of the relationship between logic (the domain of mathematics) and theory. He explains further:

> Which, then, is cartography, theory or logic? . . . If we say a^2 plus b^2 equals c^2 without identifying the a, b, and c with observable phenomena, then we are dealing with pure mathematics, a system of logic, a deduced system of relationships. But if we identify the a, b, and c with some observable phenomena, say, numbers of apples, oranges and peaches, then the formula is a theoretical statement. In similar fashion it might be argued that the map is capable of portraying the spatial property shape without reference to any observable shape. But if we map the outline of Long Island, we are dealing with theory, since we have identified the abstraction shape with a particular set of observable facts, the outline of Long Island.[14]

At the dawn of the 1960s, Bunge's work provides a useful mile marker. Certainly, the role of the map was integral – not only as representation of geographic phenomena or illustration of regional differentiation but as a kind of logic. In so doing, Bunge suggests that the map would need to become the vehicle for geography's rise to the level of science.

The experiments with computer-based cartography by Bunge and these Washington graduate students, allowed them to establish a kind of logic to geographic theory and observable facts. Alongside these experiments and conceptualizations came innovations in computer mapping. For instance, in a paper from 1959, Waldo Tobler explains a method of using 343 punch cards to plot a map of the contiguous United States in a lightning-fast 15 minutes (Tobler 1959). They were

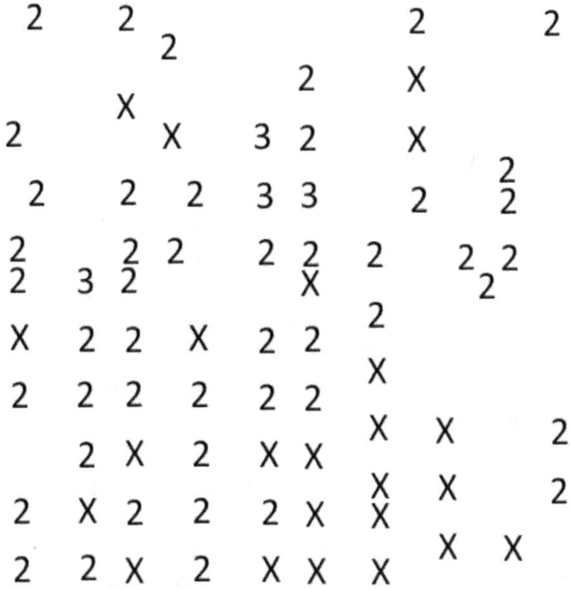

Figure 5.1 Enlarged and redrawn section of Ed Horwood's 1962 digital map of urban blight in Spokane by census block.

Source: Created by author

learning these methods because many of the grad students had funding to work in the Washington civil engineering department, under Edgar M. Horwood (1919–85) who was awash in grant money.

Horwood was exploring computational methods to study urban blight. Figure 5.1 demonstrates the Horwood method, created with the energy and innovations of the Garrison Raiders. Such a field of numbers represents blight by census block in Spokane, Washington, and is, in my opinion, one of the first leaps in the rise of the digital map.[15] Horwood traveled the country in those early years of the 1960s, sharing this computer-mapping method. One such workshop in the summer of 1963 was held at Northwestern University, on the topic of computer applications for urban analysis. At this workshop, Horwood met Howard Fisher. Fisher, impressed by the possibility of this method, but not of its readability, took to innovating on the approach.

SYMAP

In early 1964, Betty Benson worked with Howard Fisher at Northwestern to program SYMAP, even giving it its nickname.[16] SYMAP employed a method of overprinting symbolism that would make Horwood's printouts easier to comprehend. A press release, "Computer Mapping Technique Developed at

Northwestern," was released on December 9, 1964, no doubt amid concerns that the origin story of the system would get muddled by Fisher's move to Harvard:

> Fisher said the process may prove useful to city officials, planners, urban and real estate analysts, or businessmen needing maps – especially maps which are expensive or complicated when drawn by hand. . . . 'A map which might cost $200 to draw by hand could cost as little as $10 to do by computer', Fisher said. Its production time could be reduced from a week or more to a few hours, he said, with actual computer time amounting to no more than a minute or two. . . . Using SYMAP, Fisher said it is possible to produce complex maps which are not only geographically more accurate, but also illustrate such things as residential housing conditions in an area, assessed valuations of real estate, crime statistics, health statistics, market information, and a wide variety of social and economic data.[17]

Figure 5.2 demonstrates the use of this overprinting process, what former users of the program might recall as OXAV.[18] While there were more advanced computing machines and output options, Fisher's vision was committed to using machinery that made digital map "mass production at different computing centers" possible nearly "anywhere."[19] Densities and surfaces were produced in ways which could be read by a novice map reader.

After visiting in 1964, Fisher arrived permanently at Harvard and formed the Laboratory for Computer Graphics and made a proposal to the Ford Foundation in September 1965 for which Harvard was awarded nearly $300,000. The stories that surround the lab typically center in on such figures as Jack

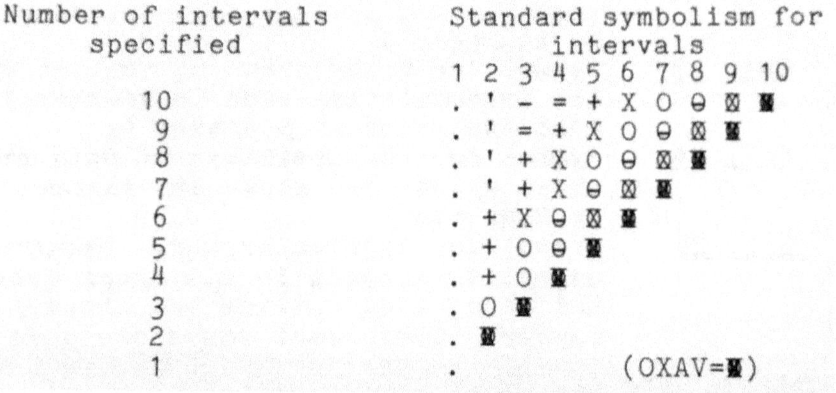

Number of intervals specified	Standard symbolism for intervals									
	1	2	3	4	5	6	7	8	9	10
10	.	'	-	=	+	X	O	⊖	⊠	■
9	.	'	=	+	X	O	⊖	⊠	■	
8	.	'	+	X	O	⊖	⊠	■		
7	.	'	+	X	⊖	⊠	■			
6	.	+	X	⊖	⊠	■				
5	.	+	O	⊖	■					
4	.	+	O	■						
3	.	O	■							
2	.	■								
1	.									

(OXAV=■)

Figure 5.2 OXAV symbolism for SYMAP data class intervals, in the 1975 SYMAP User Reference Manual by James A. Dougenik and David E. Sheehan. Harvard University Laboratory for Computer Graphics and Spatial Analysis.

Source: Image courtesy of Harvard University Archives

Dangermond, who was recruited by Fisher and who earned his MLA at Harvard in 1969 before forming Environmental Systems Research Institute (Esri) with his wife, Laura, in Redlands, California. In the early days of his consultancy, Dangermond used and developed the SYMAP program further, and he would go on to recruit members of the lab to join operations at Esri.

William Warntz joined in 1966. Previously at the American Geographical Society in New York City with training as an economic geographer at Penn, Warntz had worked with John Q. Stewart at Princeton to advance a spatial social physics. Warntz was named professor of theoretical geography, although he did not hold a chair at Harvard – a significant distinction. He was the first professor of geography at Harvard since the death of Whittlesey, 10 years prior. Allan Schmidt (1935) was recruited to the lab by Fisher in 1967. Schmidt was at Michigan State working on a program called METROPOLIS, a kind of urban simulation routine. He had used SYMAP to create an animated cartography of Lansing, Michigan.[20] Fisher put him in place as associate director of the lab, which, with Warntz, became the Laboratory for Computer Graphics and Spatial Analysis.

Fisher's practice of correspondence was unrelenting, placing him in touch with many of the contemporary thinkers in academic geography, including Robinson at the University of Wisconsin. In a letter to Robinson in 1966, Fisher writes, "We are primarily concerned with the goal of increasing the communication value of the type of maps we are producing."[21] The "communication value" of SYMAP (shown in Figure 5.3[22]) was likely in question by Robinson and other cartographers of the day, such as George Jenks at the University of Kansas. However, Fisher felt that in time these prickly issues would

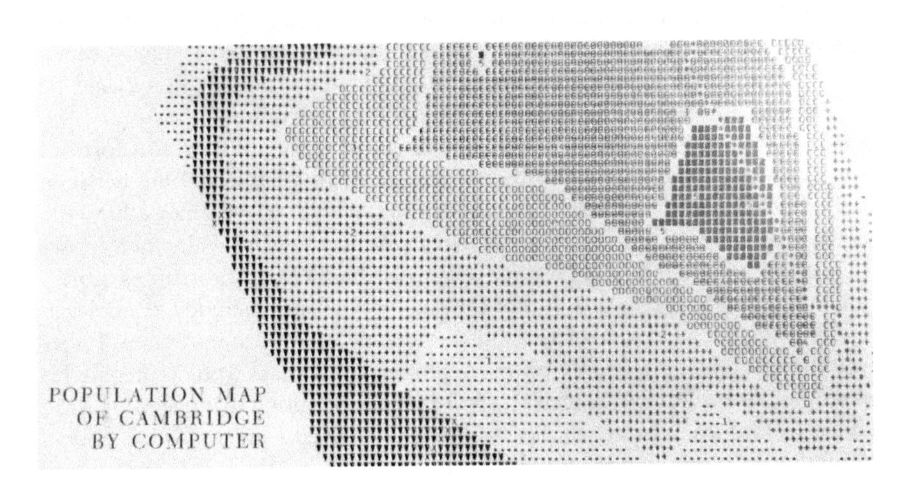

Figure 5.3 Section of SYMAP output showing population density in Cambridge, created in 1966 by James A. Dougenik and David E. Sheehan. Harvard University Laboratory for Computer Graphics and Spatial Analysis.

Source: Image courtesy of Harvard University Archives

fade – that with greater experimentation with and exposure to computer mapping methods, cartographers would broaden and deepen their impact. For him, decisions about the symbols used to represent a statistical surface were often left to aesthetics, instead of a more data-led method. At the 1970 conference at Northwestern, David Sheehan (a member of the lab) gets into an argument with Robinson:

Mr. Sheehan: As I said, I think geographers tend to look towards map aesthetics more than they do towards data.

Professor Robinson: I think what I want to object to is the term "aesthetics." There is a tremendous amount of information in a dot map because the dot map by itself is countable. The densities are not easily calculated in one's head, but that doesn't take away the fact that there is a tremendous locational value in a dot map.

Mr. Sheehan: As I said, otherwise, I think it is time that dot maps graduated, so to speak. They are at, basically, a kindergarten stage, . . . I think it is time that dot maps graduated from high school and went to college.

Professor Robinson: I think they are in the kindergarten stage, and I think they ought to stay there. I would be very happy, indeed, if all dot maps that I have seen wouldn't have any legend at all on them. . . . Being scientists, geographers, planners and so on, you have a tendency to want data whether you need it or not.[23]

This exchange highlights the uneasiness, but the serious evaluation, of thematic mapping with computing machinery. The computer provided a much-needed interruption in the field, as decisions about surface representation could be rationalized, isolated, and brought into greater comparison.

Beyond the debates about map aesthetics and the role of data, this form of map production was a curiosity to cartographers. In a letter in 1966 between Robert Williams of the Yale Map Collection and Fisher, we can see how the method of interpolation is problematized. Williams writes, "In many cases the pattern obtained by the contour option of the SYMAP program is more a function of the program than it is of the data."[24] Williams included examples of SYMAP output to demonstrate the heavy-handedness of the program. Fisher would often annotate directly on correspondence, perhaps to help prepare his response. He circles and corrects the word program, stating that instead it is the "user's lack of knowledge" that causes such ill-derived patterns. Fisher understood that maps are imperfect representations. Williams also raised the issue of responsibility, given the persuasiveness of maps: "Since maps are perhaps the most persuasive form of communication, those who make them must accept an unusually high degree of responsibility for their truthfulness. This responsibility is increased when the aura of infallibility of the computer is added to the map."[25]

Using his red pen, Fisher underlined truthfulness, drawing a line to the margins of the letter, and wrote, "all maps are a con." Fisher understood that the computer disrupted the relationship between the author of the map and the reader, a relationship that had always been based largely on deception.

SYMAP was originally developed on an IBM 709 with a deck of around 3000 cards. Amid many skeptics, Fisher felt a need to carefully innovate and demonstrate the significance of computing power for mapping. In a letter assuring a scholar at McGill, Fisher writes,

> I understand fully your skepticism regarding the computer but that may well be because people tend to exaggerate what it can do. It is no substitute for human judgment or creative imagination, intuition, etc. But it is an invaluable tool that can be of vast aid.[26]

A correspondence course in the use of SYMAP enrolled more than 500 participants in 1967. An innovation itself in terms of cartographic pedagogy, Fisher and his team designed a method by which computer mapping could be taught to the masses. Participants would complete a coding form, using a SYMAP ruler, and then mail the completed coding form to the lab, where that form was used to create a series of computer punch cards. These would then be used on the computer to produce an output that would be mailed back to the participant. As a result, participants learned the basics of map techniques supported by SYMAP, and, perhaps more importantly, they learned "how one typically communicates with a computer, and something of the types of information that a computer requires to operate."[27] Undoubtedly, the dedication of the lab staff to ensuring this correspondence course's success would greatly expand the reach of digital mapping techniques.

Metadata

By 1970 Fisher was officially retired, maintaining a research title at Harvard without much salary. He was devoting much time to the book project, along with a growing interest in color models – with additional funding from the Ford Foundation to finish the writing. In this letter to Don Shepard in June 1970, his freshmen seminar student from 1966, he writes somberly, "It appears that we were just a little ahead of our time. That's been the story of much of my life, and I don't recommend it."[28] Society seemed incredibly delicate; problems were abundant. Geoff Dutton, a student researcher under Warntz, recalls the setting of the lab in the basement of Memorial Hall:

> Olive drab canisters of crackers and drinking water were stacked in there, remnants of the fallout shelter craze of the late 1950's. This scene now seems a metaphor for how the Lab sheltered me from the fallout of academic politics and campus unrest as the sixties turned to the seventies.[29]

College campuses across the United States were erupting in protest. Harvard was no exception, and many members of the lab were embroiled.

Fisher strongly believed that computer mapping was a solution of great potential, that with greater mapping would come greater data collection and better comprehension of global and local problems. In some writing that likely were preparatory for his book project, he clarifies,

> For – just as the purpose of computing is insight, not numbers – so the goal of mapping is understanding, not maps. To a very significant degree the problems of future war and peace may be favorably affected by the better knowledge to be achieved through the improved mapping of better data over coming years. It is, we understand, the position of various students of the Vietnam experience that, had there been a better understanding of the spatially variable facts in and surrounding Vietnam, the war would almost certainly never have been blundered into by either side.[30]

Had more information been made more comprehensible, he believed, "we probably would not have gotten into the war."[31] Fisher continued to connect the Vietnam War with urban management:

> To the extent we are necessarily concerned today with difficulties to be found in our large and rapidly growing cities, not only in the United States but in most countries of the world, the problem is of a similar nature. The difficulties of our ghettos, for example, involve as a necessary and inherent fact spatially variable information in and about the ghetto. . . . For an adequate understanding of ghetto problems in relation to the city as a whole . . . it is clear that an adequate knowledge of the spatially variability of pertinent information is crucial. As in the case of Vietnam, the facts of value that are available today far exceed the facts which are being used. In policy making and in other connections, we are not employing effectively more than a fraction of the information that is available to us – and we are not sufficiently encouraging the collection of still better and more valuable data as needed.[32]

While the Fisher papers do not clearly demonstrate nor explain the involvement of the lab in the war effort, correspondence between Fisher and Tom Thayer, the deputy director of intelligence and force effectiveness of the Office of the Assistant Secretary of Defense, in March 1967 details the needs of the National Military Command Center for a "graphic display system" to analyze data sources to include friendly and enemy forces, enemy incidents, and friendly casualties, toward the "integrated analysis of the situation in Southeast Asia."[33] The story of the lab's specific role in the computer mapping of the Vietnam War is emerging (Belcher 2019).

In a memo to Carl Steinitz in 1967, Fisher encouraged working with Daniel Conway, a student of Carl Rogers from the University of Pittsburg, who was a student of Horwood at Washington.[34] These lineages mattered, in a time of sweeping change. However, in his later years, Fisher wanted to establish where Horwood left off and SYMAP began. In 1975, he wrote to Horwood to recover that particular history, "In a word while the SYMAP program was not an outgrowth of your program – in the sense that it was not built upon it – it was entirely the result of the stimulating experience which you provided to me."[35] Later in 1975, in a memo to Allan Schmidt, he clarifies,

> There is no question that Ed's course led to its development – but as a form of rebellion against his so-called maps which were merely numbers printed on plain white paper. . . . Ed never did produce by computer anything that could be called a map – at least previous to the development of SYMAP.[36]

In 1975 and 1976, much of Fisher's correspondence sought to thank those involved in the earliest days of the lab (including Betty Benson, Waldo Tobler, and Brian Berry).

The lab was in peril during much of the 1970s. Bill Warntz resigned in a huff in 1971.[37] The GSD considered closing the lab in 1974, which coincided with the death of Bob Weinberg. Weinberg was a GSD alumni and friend of Fisher who had helped support the formation of the lab before Fisher won the Ford Foundation grant. Fisher quickly negotiated a donation from the Weinberg trust to the GSD as a shot in the arm. A Computer Graphics Prize was quickly established, and Brian Berry, another Garrison geography grad, was parachuted in from the University of Chicago to direct the lab. By 1978, software from the lab was in operation in 26 countries outside the United States. The lab was experimenting with new techniques and new software, with rapidly evolving computing machinery and display technologies.[38]

Howard Fisher died in January 1979 at the age of 75, leaving much of his work on color models unfinished, as well as the book project funded by Ford. His friends and colleagues set about producing a text based on his manuscripts, led by Allan Schmidt, and a local publisher was found to release the book in 1982: "Mapping Information: The Graphic Display of Quantitative Information." The book emerged in a field that was rapidly changing. It offered little in the way of the history of computer mapping, and instead focused on Fisher's experimentation with classes and generalization. Few seemed to pay attention. Robinson's fifth edition of "Elements of Cartography" would emerge in 1984, largely unscathed by the lab's work.

Fisher recognized the power of maps long before cartography would turn critical. Mapping, then as now, was more than a tool. For Fisher, map applicability was deeply connected to comprehension, even psychology. As we return to and reconsider the digital methods developed at the lab, that experimental spirit remains – that the line of comprehension should not limit mapmakers but should instead inspire them toward new ends. This is perhaps Harvard's most important contribution to the practice of 20th-century geography – a contribution born precisely in the context of a failed academic discipline at Harvard. The year 1948, therefore, is not only a moment of continental disappointment but perhaps a moment that also "cleared the deck" for something else, something new, quite vulnerable, and risky, that might actually change the game.

Notes

1 This chapter is an abbreviated version of Wilson (2017): Digitality: Origins, or the Stories We Tell Ourselves" in *New Lines: Critical GIS and the Trouble of the Map* (Minneapolis: University of Minnesota Press, 2017).
2 Schaefer (1953) had attempted in a critique of Richard Hartshorne, the author of *The Nature of Geography,* in 1939 by establishing the role of methodology.
3 I am grateful to Nick Chrisman, a former mentor of mine at the University of Washington, and his autobiographical account of the LCGSA (see Chrisman 2006: *Charting the Unknown: How Computer Mapping at Harvard Became GIS*).
4 A full history of the social reproduction that produced the digital map has yet to be written, although work by Judith Tyner (1999) and Will van den Hoonaard (2013) are notable and needed interruptions.
5 It is shameful that very little has been written or documented about Betty Benson (born on October 6, 1924, and died on October 15, 2008), one of many women enrolled in the work of the LCGSA in the 1960s and 1970s. Her obituary notes that she held a master's degree in anthropology from Northwestern and worked in Evanston as a computer programmer; see "Death Notice: Betty Tufvander Benson," *Chicago Tribune,* October 19, 2008.
6 Transcription from Northwestern University Conference (Fisher 1970a).
7 A reproduction of this agreement, dated June 18, 1930, is available for viewing at the Recording Secretary's Office of Alumni Affairs and Development at Harvard University.
8 Raisz had worked with Davis's student Douglas Johnson at Columbia University, while completing his doctorate in geology (see Robinson 1970).
9 Raisz 1938, *General Cartography*.
10 See also the Raisz experiments with the "armadillo projection" (Raisz 1943).
11 Actually, a department of geography was never fully established at Harvard. Conant's remarks ended an ongoing discussion about the creation of a department.
12 Robinson details in Erwin Raisz's obituary that following the demise of the institute, Raisz held teaching assignments at the University of Virginia, University of Florida, University of British Columbia, and Clark University (Robinson 1970; see also Ackerman 1957; Harvard Crimson 1956a, 1956b).
13 Personal communication with Dick Morrill, February 7, 2014. (See also Entrikin and Brunn 1989, 76–80.)
14 Ibid. 26, emphasis original.
15 Spokane Blight by Census Block, Box 11, Accession Number 3365–87–13, Papers of Ed Horwood, Special Collections, University of Washington Libraries, Seattle, Wash.
16 The SYMAP User's Reference Manual, fifth edition, second printing (October 1975), contains a two-page history of the program, establishing the key break from Horwood's Card Mapping Program. Fisher corresponded with Horwood in October and November 1975 on the topic of crediting Horwood's workshop in August 1963. In addition to Betty Benson (at Northwestern), other programmers mentioned included those at Harvard: Robert Russell, Donald Shepard, Marion Manos, and Kathleen Reine (see *SYMAP User's Reference Manual*, Fisher [n.d.]a, ii–iii).
17 Press Release from Northwestern, December 9 (Fisher 1964).
18 SYMAP User's Reference Manual (Fisher [n.d.]a, Sections 3, 10).
19 Howard Fisher to David K. Camfield, May 11 (Fisher 1966a).
20 Schmidt was working with Richard Duke at MSU, also with funding from Ford. Schmidt recalls meeting Fisher at an American Institute of Planning conference. At the time, Schmidt was working in Louisville, Kentucky, where he completed the SYMAP correspondence course.

21 Howard Fisher to Arthur Robinson (Fisher 1966b).
22 Fisher submitted a portfolio in November 1973 for an American Institute of Architects Fellowship. See SYMAP example (Fisher [n.d.]c).
23 Transcription from Northwestern University Conference (Fisher 1970a).
24 Robert Williams to Howard Fisher, March 22 (Fisher 1966d).
25 Williams to Fisher, March 22 (Fisher 1966d).
26 Howard Fisher to John Bland, May 13 (Fisher 1966c).
27 Summary of Introductory Correspondence Course (Fisher 1967c).
28 Howard Fisher to Donald S. Shepard, June 16 (Fisher 1970b).
29 Geoff Dutton, Memoir, May 5, 1966, prepared for the occasion of the dedication of the Harvard Center for Geographic Analysis, shared with the author.
30 Notes on the book project (Fisher [n.d.]b).
31 "Computers Can Draw Maps, and Public Works Men Can Use Them," *APWA Reporter*, July 1970 (Fisher 1970a).
32 Notes on the book project (Fisher [n.d.]b).
33 Tom Thayer is a central figure in a discussion of the Hamlet Evaluation System (HES) (see Kalvas and Kocher 2009; Thayer 1985). For the single, conspicuous piece of correspondence with Thayer in the Fisher papers, see Thomas M. Thayer to Howard Fisher, March 1 (Fisher 1967a). Here, Thayer indicates that Robert Taylor at the Advanced Research Projects Agency (ARPA, now DARPA) would be brought in to discuss the funding of Fisher's research. Taylor was tasked to set up a computer at the Military Assistance Command Vietnam (MACV); see Robert Taylor, "An Interview with Robert Taylor," by William Aspray, *The Center for the History of Information Processing,* February 28, 1989, Charles Babbage Institute, University of Minnesota. My own conversations with Carl Steinitz indicated that members of the lab were certainly contacted regarding computer mapping techniques in support of the Department of Defense efforts in Vietnam.
34 Howard Fisher to Carl Steinitz, March 17 (Fisher 1967b).
35 Howard Fisher to Ed Horwood, October 13 (Fisher 1975a).
36 Howard Fisher to Allan Schmidt (Fisher 1975b).
37 Howard Fisher to William Warntz, August 10 (Fisher 1971).
38 The lab would eventually become embroiled in issues of software royalties and the use of the Harvard brand on lab materials (see Chrisman 2006).

References

Ackerman, E. A. (1957): Derwent Stainthrope Whittlesey. In *Geographical Review*, 47(3), 443–445.

Barnes, T. J. (2001): Lives Lived and Lives Told: Biographies of Geography's Quantitative Revolution. In *Environment and Planning D: Society and Space*, 19(4), 409–429.

Belcher, O. (2019): Sensing, Territory, Population: Computation, Embodied Sensors, and Hamlet Control in the Vietnam War. In *Security Dialogue*, 50(5), 416–436.

Bunge, W. (1968): Fred K. Schaefer and the Science of Geography. In *Harvard Papers in Theoretical Geography: Special Papers Series, n.A*, 1 November: 1–22.

Chicago Tribune (2008): Death Notice: Betty Tufvander Benson. 19 October.

Chrisman, N. R. (2006): *Charting the Unknown: How Computer Mapping at Harvard Became GIS*. Redlands, CA: ESRI.

DeVivo, M. S. (2014): *Leadership in American Academic Geography: The Twentieth Century*. Lanham/Boulder/New York/London: Lexington Books.

Entrikin, J. N.; Brunn, S. D. (1989): *Reflections on Richard Hartshorne's the Nature of Geography, Occasional Publications of the Association of American Geographers*. Washington, DC: Association of American Geographers.

Fisher, H. (1964): *Papers of Howard T. Fisher.* Harvard University Archives, Box 16. Northwestern University Folder, HUGFP 62.7, Cambridge, MA.

Fisher, H. (1966a): *Papers of Howard T. Fisher.* Harvard University Archives, Box 20. Southern Illinois University Folder, HUGFP 62.7, Cambridge, MA.

Fisher, H. (1966b): *Papers of Howard T. Fisher.* Harvard University Archives, Box 23. "Wisconsin, University of" Folder, HUGFP 62.7, Cambridge, MA.

Fisher, H. (1966c): *Papers of Howard T. Fisher.* Harvard University Archives, Box 15. N-Correspondence Folder, HUGFP 62.7, Cambridge, MA.

Fisher, H. (1966d): *Papers of Howard T. Fisher.* Harvard University Archives, Box 23. Williams Robert Folder, HUGFP 62.7, Cambridge, MA.

Fisher, H. (1967a): *Papers of Howard T. Fisher.* Harvard University Archives, Box 22. T-Correspondence Folder, HUGFP 62.7, Cambridge, MA.

Fisher, H. (1967b): *Papers of Howard T. Fisher.* Harvard University Archives, Box 20. Steinitz Carl F. Folder, HUGFP 62.7, Cambridge, MA.

Fisher, H. (1967c): *Papers of Howard T. Fisher.* Harvard University Archives, Box 21. SYMAP Correspondence Course Materials Folder 1, HUGFP 62.7, Cambridge, MA.

Fisher, H. (1970a): *Papers of Howard T. Fisher.* Harvard University Archives, Box 15. Northwestern University Conference, Miscellaneous Correspondence, Memoranda and other Papers Folder 2, HUGFP 62.7, Cambridge, MA.

Fisher, H. (1970b): *Papers of Howard T. Fisher.* Harvard University Archives, Box 20. Shepard Donald S. Folder, HUGFP 62.7, Cambridge, MA.

Fisher, H. (1971): *Papers of Howard T. Fisher.* Harvard University Archives, Box 23. W-Correspondence Folder, HUGFP 62.7, Cambridge, MA.

Fisher, H. (1975a): *Papers of Howard T. Fisher.* Harvard University Archives, Box 14. Misc. Correspondence 1975–1977 Folder 1, HUGFP 62.7, Cambridge, MA.

Fisher, H. (1975b): *Papers of Howard T. Fisher.* Harvard University Archives, Box 14. Misc. Correspondence 1975–1977 Folder 2, HUGFP 62.7, Cambridge, MA.

Fisher, H. (n.d.a): *Papers of Howard T. Fisher: SYMAP User's Reference Manual.* Harvard University Archives, Box 22. HUGFP 62.7, Cambridge, MA.

Fisher, H. (n.d.b): *Papers of Howard T. Fisher.* Harvard University Archives, Box 20. Significance, Scope, and Organization Folder, HUGFP 62.7, Cambridge, MA.

Fisher, H. (n.d.c): *Papers of Howard T. Fisher.* Harvard University Archives, Box 18. Portfolio for AIA Fellowship Nomination 1 Folder, Papers of Howard T. Fisher, HUGFP 62.7, Cambridge, MA.

Harvard Crimson (1956a): Geography at Harvard. 26 November.

Harvard Crimson (1956b): Well-Known Geographer Derwent Whittlesey Dies. 26 November.

Kalvas, S. N.; Kocher, M. A. (2009): The Dynamics of Violence in Vietnam: An Analysis of the Hamlet Evaluation System (HES). In *Journal of Peace Research*, 46(3), 335–355.

Morrill, R. L. (1984): Recollections of the "Quantitative Revolution's" Early Years: The University of Washington 1955–65. In Billinge, M. (Ed.): *Recollections of a Revolution.* London: MacMillan, 57–72.

Raisz, E. J. (1938): *General Cartography.* New York: McGraw-Hill, vii.

Raisz, E. J. (1943): Orthoapsidal World Maps. In *Geographical Review*, 33(1), 132–134.

Robinson, A. H. (1970): Erwin Josephus Raisz, 1893–1968. In *Annals of the Association of American Geographers*, 60(1), 189–193.

Schaefer, F. K. (1953): Exceptionalism in Geography: A Methodological Examination. In *Annals of the Association of American Geographers*, 43(3), 226–249.

Smith, N. (1987): "Academic War Over the Field of Geography": The Elimination of Geography at Harvard, 1947–1951. In *Annals of the Association of American Geographers*, 77(2), 155–172.

Thayer, T. C. (1985): *War without Fronts: The American Experience in Vietnam, Westview Special Studies in Military Affairs*. Boulder, CO: Westview.
Tobler, W. R. (1959): Automation and Cartography. In *Geographical Review*, 49(4), 526–534.
Tyner, J. (1999): Millie the Mapper and Beyond: The Role of Women in Cartography since World War II. In *Meridian*, 15, 23–28.
van den Hoonaard, W. C. (2013): *Map Worlds: A History of Women in Cartography*. Waterloo, Ontario: Wilfrid Laurier University Press.

6 Multivariate functions

Heterogeneous realities of quantitative geography in Hungary

Ferenc Gyuris

Abstract

Although the term "quantitative revolution" is widely used to mark specific times and events in the history of geography, it is difficult, if not impossible, to precisely define its meaning. The analysis of quantitative data, the application of sophisticated statistical methods, the construction of mathematical models and formulas, the use of a geometrized geodesign, the willingness to reveal universal laws that can be utilized in the solution of practical problems, and the identification of the self as a "quantitative geographer" are but a few of the features that may make up an idealistically conceived quantitative geography. In my paper, I will trace the emergence of these features in geographical works over the history of Hungarian geography. In doing so, I will develop the argument that the diversity of scholars and institutions involved in quantitative research goes far beyond what "clear-cut" narratives in mainstream Anglophone geography may imply.

Some early examples of quantitatively analyzing spatial issues were already present in early 20th century and interwar Hungarian geography. Then, the radical transformation of the discipline in the early Communist period opened the floodgates to attempts to make geography an application-oriented discipline without the use of sophisticated quantitative methods. After the process of destalinization, due to a complex set of changes in the political, social, economic, and academic framework, some Hungarian scholars produced a considerable number of archetypical works in quantitative geography, and the tradition has been present since then. However, such scholars have conceived quantification in different ways. They have not belonged to a single scholarly circle in geography. They have worked at diverse places, had different motivations and international and interdisciplinary connections, and even identified themselves in different ways; they have been present in various domains of geography and spatial sciences more broadly, but never actually dominated the whole of the discipline.

DOI: 10.4324/9781003122104-6

Antecedents of quantification, modeling, and applied approaches in Hungarian geography

Looking for general spatial regularities, creating models with general relevance, using "sophisticated" mathematical and statistical methods as well as geometrized cartographic and illustrative designs, and even the willingness to promote "modern" and "application-oriented" research, all had some antecedents in Hungarian geography before the wave of quantification in the 1960s. These issues return to the previous history of Hungarian geography. More than four decades after the opening of the country's first university department of geography in 1870 in Pest (today Budapest) and the founding of the Hungarian Geographical Society in 1872, World War I and the Trianon Peace Treaty in 1920 brought about massive changes in the discipline. Hungary surrendered roughly two-thirds of its territory, more than half of its population, and approximately one-third of those whose native language was Hungarian. Therefore, territorial revision, or reclaiming the surrendered areas, became an ultimate goal for the country's interwar national-conservative elite. This resulted in an increasing reputation of geography as a discipline that was witnessing remarkably improving personal, institutional, and financial opportunities. Many representatives of the Hungarian scientific and political elite presumed that successful territorial revision requires well-elaborated arguments, which are in concert with state-of-the-art international theoretical concepts and methodologies and thus can successfully be employed during negotiations with French, British, and US diplomats (Gyuris 2014a; Győri and Gyuris 2012, 2015; Győri and Withers 2019). As a result, Hungarian geography soon became internationally up-to-date in the most important domains.

These improvements opened the door to a much broader application of statistics than before.[1] The first boom in using statistics was related to ethnic mapping in service of preparations for post-WWI peace negotiations and barely went beyond presenting data on social distribution by ethnicity, language, and religion, and eventually the calculation of some absolute volumes and simple percentage values (Gyuris 2009). This soon started to change, however. Quantitative methods emerged mainly in two scholarly circles around Pál Teleki and Gyula Prinz. Teleki (1879–1941) was one of the most influential geographers and politicians in Hungary, and he also served as the prime minister of the country for two terms (1920–1921 and 1938–1941). He established the Faculty of Economics at the University of Budapest and he also chaired its newly opened Department of Economic Geography, which soon became a prominent school of human geography (Ablonczy 2007; Győri and Gyuris 2012, 2015). Teleki's approach was based on Vidalian *géographie humaine*, which he introduced into Hungarian geography in his seminal 1917 work "The History of Geographical Thought" (Teleki 1917). He also took many elements from contemporary

British, American, and German geography. This diverse international influence can also be identified in the works of his coworkers and disciples, with individually different foci. Gyula Prinz (1882–1973) had a different background. He studied at the universities of Budapest, Berlin, Munich, and Breslau (today's Wrocław, Poland), from which he returned to Budapest in 1904 with a degree already in hand. His scholarly work stretched from geology to human geography and showed a firm influence of contemporary German geography, even during Prinz's consecutive professorships at the University of Pécs (1923–1940) and Kolozsvár (today Cluj-Napoca, Romania) (1940–1944), which Hungary regained from Romania for a few years in 1940 (Fodor 2006; Jobbitt and Győri forthcoming).

Quantitative attempts mainly took place in (1) political, economic, and urban geography, where the representatives of the "Teleki school" were especially active, as well as in (2) administrative and transport geography, where Prinz and his disciples proved more prolific. The main objective in political geography was to prepare for the envisaged territorial revision of national boundaries in Europe, which essentially meant the fabrication of scientific arguments for the expected new international negotiations. In 1926, Teleki established the *Államtudományi Intézet* (Institute of Political Science) and headed it for 13 years to continuously collect, analyze, and visualize the most detailed, precise, and state-of-the-art statistics from Hungary and neighboring countries (Rónai 1989). András Rónai (1906–1991), his young disciple and follower who became the director of the institute after 1939, authored numerous works in the field (Gyuris 2009, 2014a).

Beyond radically changing state-spaces in Central Europe, the new national boundaries of 1920 dissolved the unitary economic space of prewar Hungary and created a new economic order in the whole region. Therefore, Teleki was convinced that economic sciences were to play a crucial role in the recovery of Hungary after 1920. He and his close colleague Ferenc Fodor (1882–1967) argued for an economic geography that "is not only a descriptive science, but an investigative one as well" and that has to illuminate "the relation of production and the producing human not just qualitatively, but quantitatively as well" (Fodor 1925, 202).[2] With his assistant at the Department of Economic Geography, Ferenc Koch (1901–1974), Teleki also investigated the relevance of Thünen's concept for contemporary European and global agricultural production (Teleki 1934) as well as "The rings of European culture and embourgeoisement" (Prinz and Teleki 1937), which was partly motivated by the works of US geographers Mark Jefferson (1911) and Ellsworth Huntington (1915, 1927) on the global geography of "civilization" and "culture" (Gyuris 2019).

Quantitative methods also appeared in urban geography, mainly due to Tibor Mendöl (1905–1966) and Béla Márton (1880–1967), who worked together in the 1930s at the University of Debrecen. During his university studies in Budapest, Mendöl was also a member of Eötvös József Collegium, an elite higher education institution modeled after the École Normale Superieur in Paris, of which Teleki was the curator between 1921 and 1941, and its section of geography was led by Fodor between 1923 and 1939 (Jobbitt and Győri 2016). Mendöl was, for instance, the first scholar in Hungary to introduce rank correlation analysis into human geography (Mendöl 1936), whereas Márton analyzed the spatial location of towns in Hungary with a strong commercial profile in light of their distance from the pre-1920 national boundaries (Márton 1941) (Figure 6.1).[3]

For Gyula Prinz, his aim was to rationalize the administrative system of Hungary. The old administrative boundaries from the pre-Trianon period were mainly kept because the political elite calculated that the national boundaries would be revised in the near future. Prinz, however, pragmatically argued that "the political territorial division of [post-1920] Hungary is one of the most imperfect ones in the whole World" (Prinz 1933, 71), so it has to be fit to the new national boundaries. Similar studies were conducted by Aurél Hézser (1887–1947) and Ernő Wallner (1891–1982), who both made their *habilitation* in economic geography at Prinz's department in 1923 and 1938.

These improvements in the 1920s and 1930s did not last for long, however. Hungary was one of the losers of World War II and suffered a serious loss. The years after 1945 already witnessed the country as a part of the Soviet occupation zone, and the Hungarian academia was exposed to a massive transformation

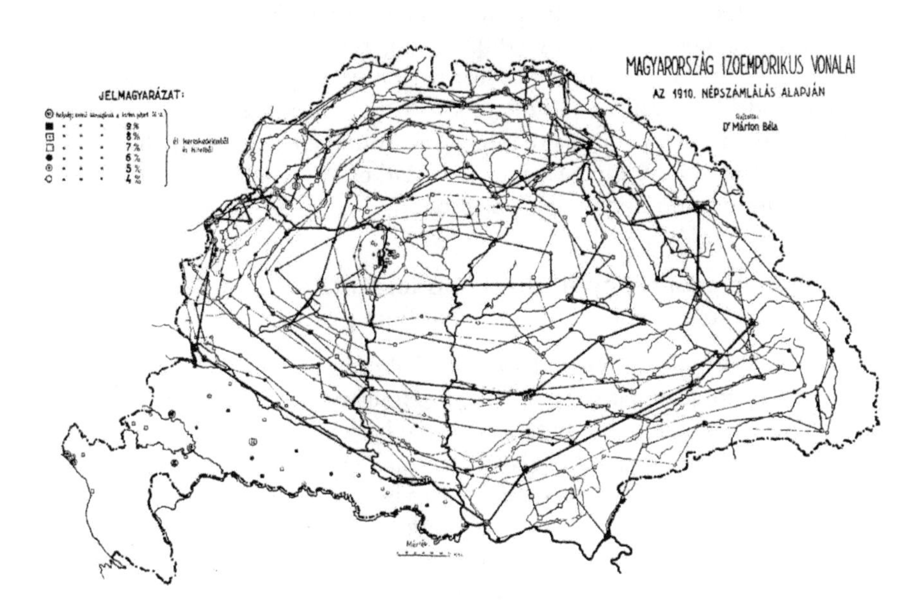

Figure 6.1 Béla Márton's 1941 map of the locations of towns with commercial profiles in pre-WWI Hungary (1910). Lines tie towns with a similar share of commerce and credit sector employees among all earners (as a percentage), proceeding along the 1910 national boundary.

Source: Márton (1941, 101); reproduced with explicit permission by the journal Földrajzi Közlemények (which published the map in 1941)

along the Stalinist Soviet model (Péteri 1998). This also resulted in the rapid emergence of a new Marxist-Leninist geography, whose representatives were explicitly aimed at a "revolutionary" transformation of the discipline. Most of the "old geography" was judged now as "guilty science" for having justified territorial revisionary aims. Leading geographers of the interwar period, as well as their most influential disciples, were mostly pushed to the periphery of academic life or forced to radically change their scientific focus (e.g., leaving human geography and focusing on politically less sensitive physical issues). Institutions of the discipline were also subjected to thorough reorganization or dissolution (Győri and Gyuris 2012, 2015).

Hungarian geography was expected to adopt the theoretical and methodical framework of its Soviet counterpart, which was continuously called "Marxist" by its representatives, even though it was built on a Marxist-Leninist conceptual basis and was, to great extent, Stalinized from the 1930s onward. *Physical geography* and *economic geography* (the new name for the domain focusing on social issues) were strongly separated to avoid any form of environmental determinism, and economic geography was now responsible for the rational allocation of population and production in space (Abella 1956). Despite this planning-oriented focus, quantitative methods were greatly peripheralized in geography during the Stalinist years, and model-based and mathematized research, again in line with academic thinking in the Stalinist Soviet Union, was regarded as "formalist." As György Markos (1902–1976), the most influential Marxist-Leninist geographer of Hungary in that period, put it, such "formalism" was aimed to hide "the content and the process behind form," and thus, especially in the social sciences, to turn attention away from the basic "contradictions of capitalism," and to serve capitalist interests (Markos 1955, 362). Yet, the attitude toward quantification started to change soon.

Quantitative geography in the 1960s and 1970s: diverse scholarly circles with diverse backgrounds

Unlike the 1950s, the 1960s and 1970s had already witnessed several valuable contributions to quantitative geography in Hungary, the authors of which did not belong to a single group with a common background. They came instead from diverse institutions, scholarly circles, and social milieus, with different individual life stories, due to which they also had diverse motivations and goals. Some of them were generally working based on pre-1945 traditions and even came from mainstream academic circles of the precommunist period. Others started from the main centers of the new Marxist-Leninist geography established after 1945, where they gained a focus on practical issues of material economic production, which they combined with newly developed personal interests in quantitative methods. In the meantime, however, both traditions had several representatives who never became involved in quantitative research.

One group of scholars producing quantitative works in Hungarian geography was related to the "Markos school" of economic geography. György Markos was already an active member of the international labor movement in the inter-war period, during which he spent most of his time as an émigré in Western countries (Austria, Germany, France, and Denmark). He returned to Hungary before World War II, writing several newspaper articles and popular books on economic issues in the following years and popularizing the idea of economic planning (Tatai 2004). In 1948, the year of the communist turn in Hungary, the Faculty of Economics at the University of Budapest, established by Pál Teleki less than 30 years earlier, became a sovereign university and was renamed in 1953 the *Marx Károly Közgazdaságtudományi Egyetem* (Karl Marx University of Economics). Although Markos did not have a formal education in geography, he was appointed the new head of the university's Department of Economic Geography. He also played a crucial role in criticizing the "old," "bourgeois" geography and its representatives (Győri and Gyuris 2012, 2015). Meanwhile, Markos was the leading promoter of the new Marxist-Leninist economic geography in Hungary after the Soviet model, which was mainly aimed at contributing to a more efficient and rational geographical distribution of the Hungarian economy (Enyedi 1976; Győri and Gyuris 2015; Czirfusz 2015). He argued for joining economic geography with spatial planning (Markos 1951). One of his disciples, Zoltán Tatai (1928–2010), called him "the organizer and propagandist of the socialist planned economy" in his recollections in 1984 (Tatai 1984, 131).

During the nine years that Markos taught at the university, he organized a scholarly group of Marxist-Leninist economic geographers. His school also educated a new generation of the discipline, the members of which landed at diverse institutions of the academic and professional sphere, including the economic geography departments at Eötvös Loránd University in Budapest (the largest in the country) and József Attila University in Szeged, the newly established (1951) Geographical Research Group (from 1967 Geographical Research Institute) of the Hungarian Academy of Sciences, and even the National Planning Bureau and the "Spatial Development" education program of the communist Political High School in Budapest (Antal 2005).

The emergence of new schools of spatial planning was backed by changes in the political framework of Hungary. In the fall of 1956, a nationwide anti-Soviet Revolution took place in the country. Although it was crushed within a few weeks and followed by a period of communist terror, the reorganized and non-Stalinist communist leadership soon launched economic reforms to increase the material well-being of the population in Hungary and decrease social unrest. The objectives of these reforms included creating a spatially more even and economically more rational allocation of the economy, which required spatial planners (cf. the 1958 government decree on regional planning and the need for regional research – Radó [1962]; some contemporary discussions of the issue are Kóródi and Kovács [1966]; Kőszegi [1967]).

Despite being a devoted Marxist, in 1955 and 1956, Markos also criticized the results of the First Five-Year Plan in Hungary (1950–1954) as well as the centralized and bureaucratic planning regime (e.g., Markos 1956a, 1956b). Hence, he was forced to leave the university department during the

Communist reprisal after the 1956 Revolution was put down, and he moved to the Geographical Research Group of HAS in 1958 (Tatai 2004). Markos did not apply sophisticated mathematics in his research, but instead argued for the extensive use of statistics and the approach of economics in economic geography (Tatai 1984).

Many disciples of Markos did not employ complex quantitative methods in their research either, and the same went for the majority of Hungarian geographers in general. Nevertheless, several scholars took a different path, the most prominent of whom was György Enyedi (1930–2012), whose 1976 obituary of Markos argued that, due to Markos's approach, "the methodological apparatus of research changed; the method of state-of-the-art economic and mathematical-statistical analysis gradually spread" (Enyedi 1976, 127) in Hungarian economic geography. Enyedi obtained his degree at Markos's department in 1953. After working there for two years, he taught at the University of Agricultural Sciences (1955–1960) before becoming a researcher of the Geographical Research Centre of HAS in 1960, up until 1983. In his early career, Enyedi focused on the geography of agriculture – an important issue both politically and economically due to the new and last large wave of forced collectivization of private land in Hungary between 1958 and 1961, followed by massive state subsidies to collectively owned agricultural production units. Moving away from a simple descriptive overview of agricultural production, which was typical in the Marxist-Leninist works of the 1950s, Enyedi – partly in cooperation with another Markos disciple, Tivadar Bernát (1926–2018), who chaired Markos's former department from 1966 to 1989 – gradually (re) introduced mathematical methods and locational theories, also referring to such attempts in interwar Hungarian geography (e.g., the works of Ferenc Fodor and Pál Teleki) and models such as those of Thünen and Lösch (Bernát and Enyedi 1961). Enyedi kept improving his concepts in an increasingly international approach. He presented his results not only in Hungarian books and journals (e.g., Enyedi 1968) but also in a 1967 paper in the US journal *Geographical Review*, which was based on a lecture he presented at the University of Chicago the year before, during his one-year research stay at UC Berkeley (Enyedi 1967). In *Földrajzi Közlemények*, the editorial report on the 1962 conference of the Hungarian Geographical Society already stressed that, "to a considerable extent," these new methods had newly become generally employed in Hungarian geography of agriculture on Enyedi's initiative (Pécsi et al. 1962, 360). Four years later, László Simon (1912–1968), then first secretary of the Hungarian Geographical Society, underscored that Enyedi's recent lecture at the Society's Division of Economic Geography was "the first to report on results achieved through the application of mathematical methods in economic geography in Hungary" (Simon 1966, 267).

A group of spatial planners, most of them from the Markos school, was also working increasingly with quantitative approaches at the National Planning Bureau, especially its Institute of Economic Planning. An important publication platform of their works was the journal *Területi Statisztika* (Regional Statistics) at the *Központi Statisztikai Hivatal* (Central Statistical Office) or KSH, which was launched in 1951 as *Statisztikai Értesítő* (Statistical Bulletin), renamed in 1956 as *Megyei és Városi Statisztikai Értesítő* (County and Urban Statistical Bulletin) before gaining its current name in 1968. *Területi Statisztika* gradually developed from a collection of official reviews into a scientific journal, providing space for a great number of regional science papers from the 1960s onward (Dusek 2010; Németh 2010).

This scholarly group was also the first in Hungary to publish an academic handbook (actually an edited volume) on "Methods of Regional Analyses" in 1976 (Kulcsár 1976). Major contributors from the Planning Bureau, who actively used quantitative methods in spatial research, included the economist and applied mathematician Csaba Csernátony, Sándor Kádas, who published a separate booklet on the international literature of regional modeling in 1976 (Kádas 1976), Gyula Wirth, an early expert of input–output models, László Francia, who wrote about the application of factor analysis in spatial research, a method he broadly used in the 1970s (Lux 2015), and László Lackó (1933–1997), who was trained as an economic geographer at Markos's department and worked for the National Planning Bureau from 1965 to 1978 before moving to the Ministry of Construction and Urban Development (Ormosy 2013). Lackó was a prominent proponent of using cartography in quantitative regional analysis and participant in the first successful experiment in Hungary with creating a cartogram with an electronic computer in 1969 (Lackó and Erdős 1970). He took advantage of the computer infrastructure of the National Planning Bureau, including an ICT 1905 computer from International Computer and Tabulators Ltd. in London (Lackó and Erdős 1970). This gave a serious advantage in a period when mathematical calculations were still seriously constrained by limited calculation capacities in Hungary (see Németh 2010). Computers only became accessible at university departments of geography in Hungary with the introduction of affordable PCs in the second half of the 1980s (see Antal and Perczel 2005; Nemes Nagy 2002), and even pocket calculators were scarce until the 1970s (Probáld 2011).

Two economists with a degree from the Karl Marx University of Economics, although not from its Department of Economic Geography, were István Bartke (1930–2009) and Iván Illés (1942–2017), who both started their career at the Ministry of Construction and Urban Development and spent a long time at the National Planning Bureau (Bartke, in the periods 1962–1971 and 1977–1990; Illés 1969–1991). They also widely applied mathematical-statistical methods in analyzing spatial issues, and Illés was the author of the first comprehensive book in Hungarian on regional economics, including the application of methods such as gravity models, Markov chains, and stochastic models of spatial expansion (Illés 1975).

At Eötvös Loránd University (ELTE) in Budapest, additional scholars became involved in quantitative methods. ELTE's Department of General Economic Geography (named the Department of Human Geography until 1950) was chaired by Tibor Mendöl for two and a half decades after 1940. Although

Mendöl did not become a quantitative geographer, he was well aware of Western quantitative works in urban geography. He analyzed in detail not only Christaller's concept but also its reputation and critique (Mendöl 1957). He remained head of the department at ELTE until 1965, although he became increasingly peripheralized in scientific power structures after the post-WWII Communist turn in Hungarian politics for being a "bourgeois geographer," and his department was gradually filled with new Marxist-Leninist scholars (Győri 2009; Gyimesi 2014). Mendöl's disciple from the pre-Communist period, Jenő Major (1922–1988), had already written his thesis in 1944 on an urban geography topic where he intensively discussed Christaller's concept (Major 1944). Major later worked at different places, including the Ministry of Construction and Urban Development, the ELTE Department of General Economic Geography, and Budapest Technical University. In 1964, he published an in-depth study, testing Christaller's concept of the urban network for the case of Hungary and drawing the hexagons of Christaller on Hungary's map (Figure 6.2). He did so neither to provide justification to Christaller's concept nor to glorify it, but rather to discuss all the methodological aspects involved in such an analysis.

Mendöl was followed as head of department between 1965 and 1993 by Zoltán Antal (1931–2011), who obtained his degree at Markos's department and started his career there as Markos's assistant. Antal was not working quantitatively, but his department at ELTE developed strong personal and institutional

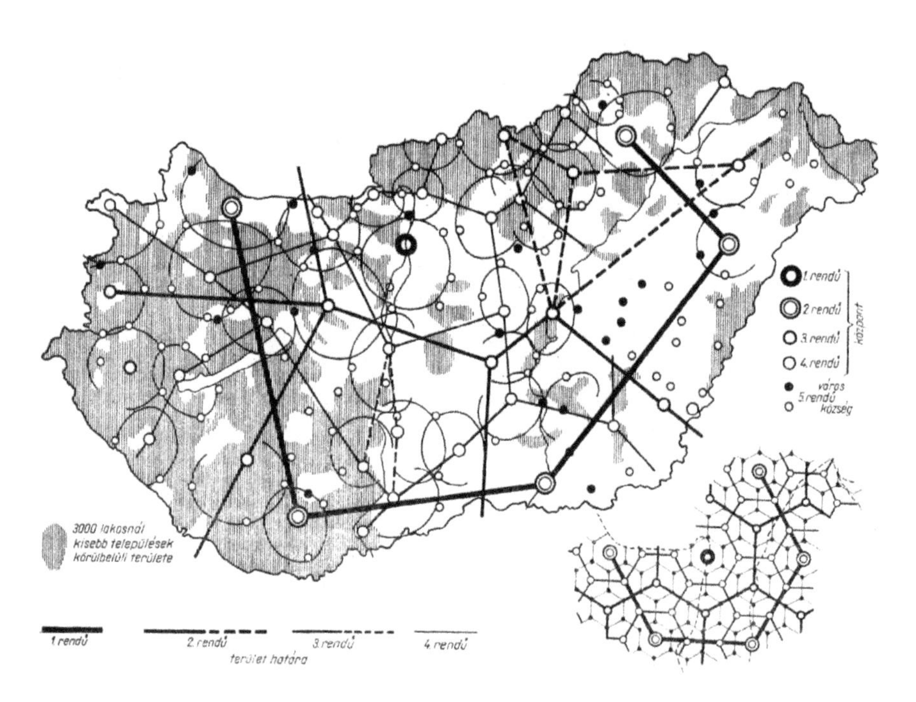

Figure 6.2 Major's map from 1964 on the spatial model of the urban network in Hungary.
Source: Major (1964, 57)

links with the National Planning Bureau, many of whose professionals regularly gave courses at the department, including Wirth, Illés, and Bartke (e.g., Antal and Perczel 2005). A member of the department's permanent staff between 1961 and 1975, Csaba Kovács (1929–2018), who also received his degree at Markos's department in 1953, published several works on the perception of space in geography and the regularities of the geographical distribution of agricultural production, including a detailed analysis of Thünen's concept (e.g., Kovács 1962, 1964, 1966). Later on, he kept working quantitatively in the National Planning Bureau after 1975.

Meanwhile, the first chair of ELTE's other nonphysical geography department, the Department of Regional Geography, was Ferenc Koch, Teleki's former assistant in the 1930s. As head of the department, Koch did not contribute to quantitative geography. His appointed successor, Gyula Dudás (1925–2000), who started to work at the department immediately after he obtained his degree in economic geography and Bulgarian language at the University of Sofia, convinced Béla Sárfalvi (1925–2000) at the Geographical Research Group of HAS, a colleague of Enyedi, to take over the department in 1967. Sárfalvi had a degree in history and geography from ELTE. He also had a high level of proficiency in English, German, and French, which he took advantage of during his one-year Ford Foundation scholarship in the US (Ann Arbor, Chicago, and Berkeley) in 1965–66 (Probáld 2002).

Ferenc Probáld (born in 1941) also joined the department in 1967. He was a biology and geography teacher and a meteorologist by training, with degrees obtained at ELTE a few years before. Initially, Probáld's focus was urban climatology, but he soon extended his research to urban social issues as well. He employed both the mathematical analytical skills and tools he internalized in the meteorology program, as well as the experience of his one-year American Council of Learned Societies (ACLS) scholarship in the US (1972–73) (Probáld 2011). His works in the 1970s included a multivariate regression analysis of the impact of air pollution on land value in Budapest (Probáld 1973) and the first use of the dissimilarity index to reveal urban segregation in Budapest (Probáld 1974). József Nemes Nagy (born in 1948) obtained his degree as a mathematics and geography teacher in the same institution in 1971, where his most influential mentor was Csaba Kovács. Nemes Nagy received his doctoral degree two years later with a more than 200-page study on "The analysis of regional differences of economic development with mathematical and statistical methods" (Nemes Nagy 1973). After that, he published several articles on related topics, both general (e.g., Nemes Nagy 1974) and with a special focus on communist countries (Nemes Nagy 1976, 1977). From 1981 onward, Nemes Nagy worked in the National Planning Bureau's Institute of Economic Planning before returning to the university department in 1994, which he then chaired for roughly two decades, until 2013.

The Ministry of Construction and Urban Development also deserves a mention in the network of scholars performing quantitative spatial analysis. Major, Illés, Bartke, and others spent several years there, which were formative for their scholarly approach. The same went for József Kóródi (born in 1930) who, with a degree from Karl Marx University of Economics (and a visiting lecturer at ELTE's Department of General Economic Geography), was the first to use a regional input–output model in Hungary (Kóródi 1959; Lux 2015).[4] These scholars gained support for their quantitative approach from the architect and urbanist Károly Perczel (1913–1992), who as a division head in the ministry played an active role in regional planning in Hungary (e.g., Perczel 1962).

In line with Hungary's centralized urban network, the key scholars of quantitative spatial research worked for institutions in Budapest. In the 1970s, an important exception was Jolán Abonyiné Palotás (born in 1942) at the University of Szeged, more precisely at its economic geography department chaired by the Markos disciple Gyula Krajkó (1929–2011). Abonyiné Palotás obtained her first degree as a teacher of biology and geography in Szeged, before getting another one in industrial economics at Karl Marx University of Economics. In collaboration with the Szeged-based mathematician Ferenc Móricz, she applied sophisticated quantitative methods, e.g., factor analysis, in several works on geographies of agriculture and infrastructure (e.g., Móricz and Abonyiné Palotás 1975).

Although having come from different schools and worked at different institutions, most of those publishing quantitative studies in the 1960s and 1970s shared having sound knowledge of mathematical and statistical methods, which they acquired during their studies in economics, mathematics (combined with geography in teacher training programs), or even meteorology. This was considered a crucial factor by several contemporaries who thus argued for improving mathematics in the university geography curricula (Kádár 1966). In addition, many of the authors of early quantitative works have explicitly emphasized in their recollections the advantage they took from such skills (e.g., Nemes Nagy 2002; Probáld 2011). Nevertheless, this seems to have been an important, although not determining factor, because several authors have stressed that they had to learn the actual and meaningful use of these methods in spatial analysis on their own, relying on literature from foreign scholars and some Hungarians in other disciplines (e.g., cartography,[5] biology, and regional economics). In addition, not all scholars with similar backgrounds performed quantitative research, whereas others who had not gained such skills during their studies at the university learned and internalized such methods in autodidactic ways (see Beluszky 2015).[6]

Diverse foreign influences

Hungary is not an isolated place, so foreign influences were important in the quantification of spatial studies. These neither came from one place, nor were driven by a single motivation. The scholars who created quantitative studies in Hungarian geography reflected the US, British, Soviet, Eastern European, or even German and French influences, which played out very unevenly on an individual scale, mainly according to a given scholar's language proficiency, personal network, and knowledge of foreign "national traditions" in geography. Although the Soviet-styled transformation of Hungarian geography cut many previous "Western" links on the individual and institutional scale, this did not result in profound isolation, not even in the Stalinist years, and the less dogmatic period of "goulash communism" from the 1960s onward provided more space for communication and knowledge exchange with scholars in Western countries through the exchange of books, people, and ideas. This took different forms, from reading, reviewing, and discussing Western literature – which also took place in scholarly journals – to guest scholarships and research stays in Western countries to attending international conferences.

In fact, mathematical analysis suddenly gained legitimacy in Soviet geography as well around 1960, in accordance with the quantitative turn in the US geography. Prominent Soviet geographers started to argue for applying mathematical methods, for example, Yulian Saushkin (1911–1982), the chair of the Department of Economic Geography of the USSR at the powerful Moscow State University from 1948 to 1981 (Saushkin 1971).[7] Soviet geographers, similar to their counterparts in other disciplines (Gerovitch 2001), also started to "de-ideologize" new Western models, arguing that they could be applied to create *objective* knowledge about the objective world, independent of the social and political system of the country in which they emerged (e.g., Zaytsev 1971). Formulas, equations, and curves were relatively easy to interpret as "value-free" and "applicable" to the Communist context. As a result, several seminal works of Anglophone quantitative geography and regional science from Walter Isard, William Bunge, Peter Haggett, Richard Chorley, and even August Lösch were published in Russian translation, just a few years after the original version, starting with 1959, through the early 1970s.[8] During the 1960s, Soviet authors discussed and argued for the use of models and mathematical analysis in several books and essays (see Jensen and Karaska 1969a; Mathieson 1969). Among more than 500 articles in leading Soviet geographical journals monitored by editors of the US translation journal *Soviet Geography*, mathematical articles were almost nonexistent in 1960 but already made up 15% of all essays by 1968 (Jensen and Karaska 1969b). In 1971, Saushkin concluded that "mathematical modeling appears to have found widest application in economic geography" (Saushkin 1971, 425); this was no longer considered "deviance" from Marxism-Leninism, but rather as "a general advance in the development of scientific thought" (Pokshishevskiy et al. 1971, 404).

These improvements also opened the gates for quantitative research in geography in other countries of the Communist Bloc. The aforementioned first academic handbook in Hungary on methods of regional analysis (Kulcsár 1976) already relied on significant literature from both "Western" and "Eastern" countries (Figure 6.3) and from a wide range of disciplines, including statistics, economics, planning studies, and geography. Similarly, Nemes Nagy (1976) could publish shortened Hungarian translations of papers employing sophisticated mathematical-statistical methods of spatial analysis (including correlation, regression, factor, and principal component analysis as well as potential models) from scholars in Czechoslovakia, Hungary, and the Soviet Union.[9]

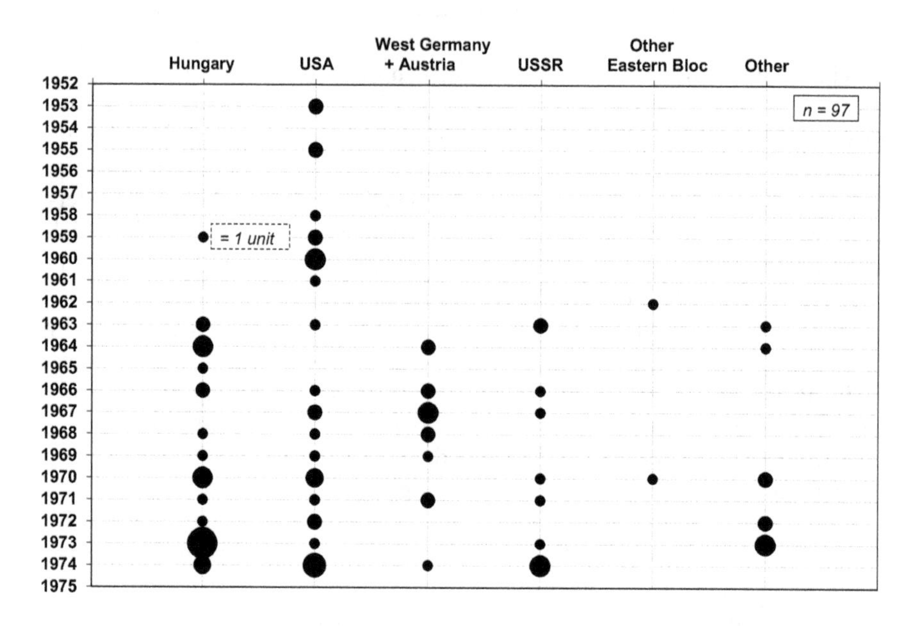

Figure 6.3 Diverse foreign influences: number of cited works in Part II (on methods and concepts) of the volume "Methods of Regional Analyses" (Kulcsár ed. 1976) by year of publication and the quoted author's country of origin (n=97). Author's own calculation and design.

Meanwhile, a decade after the International Regional Science Association held its first conference in 1954 in Detroit, its first Eastern European meeting took place in Cracow, Poland, in 1965, followed by events in Budapest in 1968 and 1975 (Isard 2003; Boyce 2004). The International Geographical Union (IGU) Commission on National Atlases also held an international conference in Budapest in 1962, which was supplemented by a geography conference of HAS and the Hungarian Geographical Society at the 90th anniversary of the latter. Here, the use of mathematical methods in economic geography was one of the main topics (Pécsi et al. 1962). The IGU also held a regional conference in Budapest in 1971, celebrating the 100th anniversary of the Hungarian Geographical Society. Quantitative analysis played a crucial role in many papers from both Western and Eastern scholars (Miklós 1972; Enyedi 1973). A bilateral Hungarian-American Geography Seminar took place in Budapest in 1975 with a focus on urban geography. The event hosted nine US scholars, including George Kish (University of Michigan, Ann Arbor), James Vance (UC Berkeley), and Richard Morrill (University of Washington, Seattle), who gave talks on mathematical-statistical methods (Mészáros 1975).

Furthermore, a few human geographers managed to spend a semester or an academic year in the Western Bloc, as was mentioned in the case of Enyedi, Sárfalvi, and Probáld. Personal experience in other Communist countries also gave a push in some cases to quantitative approaches. The 23rd International Geographical Congress in Moscow in 1976 provided much space for discussing the ongoing "thorough methodological rearmament of science" (Geraszimov et al. 1977, 297) through mathematical models, computer technology, and other new solutions. Dudás, too, who spent 10 months in the USSR in 1973 as a guest scholar in Saushkin's department in Moscow, stressed in his report after returning to Hungary, and presented as a positive example to follow in Hungarian scholarship, the widespread use of mathematical methods in the geography departments of several Soviet universities and research institutes of the USSR Academy of Sciences as well as the increasing role of mathematics in university geography curricula (Földrajzi Közlemények 1974).

Quantification after the 1980s: increasing role instead of withdrawal

From the 1980s onward, two narratives of quantification started to develop separately in Hungarian geography and regional science. A glaring example was when a group of young scholars from Enyedi's teams in two consecutive research institutes of HAS[10] published the handbook "Potentials of Applying Mathematical and Statistical Methods in Regional Research" in 1984 (Sikos 1984). This was in fact the first geography handbook in Hungary that took pains to adhere to every related statistical method and its application (Dusek 2014). Enyedi, who was not a member of the book project but wrote the book's two-page foreword, referred to the quantitative revolution as already having been "over" internationally "for 15–20 years" (Enyedi 1984, 9). He mentioned some reasons for this, concluding that "*[t]he application of state-of-the-art mathematical-statistical methods has produced barren results and remains drowned in empty formalism if those who apply it aren't even familiar in a qualitative sense with the subject on which they have collected and processed data*" (Enyedi 1984, 10). His words were in line with the postpositivist criticism in Anglophone countries during the 1980s. Tamás Sikos (born in 1953), however, the editor of the book as one of the young scholars, enthusiastically wrote about quantitative geography a few pages later as an emerging and widening domain, the further improvement of which they wanted to promote with the volume. It is remarkable to see the difference between the "senior" Enyedi and his "Western-conforming" views on the one hand, and on the other hand, the enthusiastic young scholars whose attitudes reflected their personal experience that quantitative analysis was required in the domains where they were involved, and their conviction that quantitative methods were still far less common in Hungary than they hoped. In fact, Enyedi's views were shared by a few other scholars as well, who had personal experience with Anglophone science during longer academic stays and who thus paid more attention to changes there than most of their Hungarian colleagues did.

However, unlike in the US, neither regional science nor the quantitative approach lost momentum in Hungary after the 1970s. Instead, their academic position strengthened, especially in the 2000s, mainly due to the massive regional policy mechanisms of the European Union, which Hungary joined in 2004. (The importance of regional planning had already started to increase in the second half of the 1990s, as Hungarian governments gradually fit the country's public institutional system to meet EU requirements.) This opened the door for awarding the first "doctor of sciences" (DSc) degrees in regional science (from 1995 onward), launching the first PhD program in regional science in 1997 (Rechnitzer 2009),[11] founding the *Magyar Regionális Tudományi Társaság* (Hungarian Regional Science Association) in 2002, and even renaming the Department of Regional Geography at ELTE as the Department of Regional Science in 2007 (chaired by Nemes Nagy between 1994 and 2013). Such changes in terminology indicate that the leading representatives of Hungarian regional science, and even the trained human geographers among them, mainly identify themselves not inside but outside of the discipline of geography. In fact, the Committee on Regional Science at the Hungarian Academy of Sciences, which was launched in 1986 in the Section of Earth Sciences (Section Nr. 10), moved to the Section of Economics and Law (Section Nr. 9) in 1999 (Rechnitzer 2009), whereas geography remained in Section Nr. 10.

In addition, approaches (e.g., critical and feminist) that emerged in Anglophone academia as early as the 1970s and that were critical to both intensive quantification and its underlying social, political, and economic interests gained a significant position in Hungarian geography only after the late 2000s, mainly during the 2010s. Before 1990, the communist dictatorship did not tolerate such voices in Hungarian academia. With the democratic transition in 1989–1990, the political framework underwent massive changes. Nevertheless, institutional and conceptual structures at universities and in research institutes proved very rigid and enduring even after 1990, and a generational shift in Hungarian geography only started ca. 15–20 years after the political transition (Gyuris 2018; Timár 2020). These conditions also secured more space for quantitative approaches, which had already been present for some time. Still, the academic position of regional science has remained weaker than that of geography in terms of number of scholars, international publications as well as newly issued MA and PhD degrees, and it increasingly started to face challenges in the 2010s, also for institutional and demographic reasons, i.e., the aging of its leading representatives (see Lengyel et al. 2020).

The beginning of a new discipline or an episode in geography's "organic development"? Quantitative approaches in light of individual self-identification

All the aforementioned scholars had a different view of quantification than what some "clear-cut" US narratives may imply. They have rarely written about a "quantitative revolution" in their works and recollections, and they have always placed the term in quotation marks (e.g., Probáld in Földrajzi Közlemények 1974; Enyedi 1984; Nemes Nagy 2002), positioning it as terminology widely used in "Western" (mainly Anglophone) discourses, not in Hungary. Instead, they usually referred to the new approach as the application of "quantitative" or "mathematical-statistical" methods.

For self-identification, there has been a group of scholars consciously using the adjective "quantitative" to identify themselves and rather consistently applying this approach in their scholarly works. However, they generally define themselves primarily as regional scientists or "regionalists" (singular: *regionalista* in Hungarian), not as geographers (not even those whose university degree was issued in geography). In addition, they do not claim that they wanted to "revolutionize" geography. Instead, they claimed that their aim was to create a new scientific discipline that is able to "scientifically" analyze spatial issues, unlike "traditional geography," which is rather a descriptive "general education subject" in their eyes. This view is not hard to understand, considering that several "regionalists" have not in fact been geographers by training, but rather economists and planners that became interested in the spatial perspective.

"Regionalists" have obviously played an important role in introducing the quantitative and spatial science type of mindset into geography in Hungary. Nevertheless, a considerable part of the major quantitative spatial studies published in Hungarian during the 1960s to 1980s was created by other authors, who have not identified themselves either as regionalists or quantifiers and did not consider their works as "revolutionizing" geography. Moreover, they have had numerous nonquantitative studies as well, so they have not consistently employed a quantitative approach in their research. These scholars regarded the quantitative approach as one of the manifold conceptual frameworks and methodologies that have been present in the discipline of geography and that they themselves also applied using high standards in some of their studies, usually not by itself, but in combination with other geographical concepts and methods. In other words, the emergence and expansion of a quantitative analytical framework was rather part of the "organic development" of the discipline in their view, instead of a "revolutionary moment" (see the personal recollections of Beluszky 2015; Probáld 2011).

Conclusion

Quantitative analysis gained popularity in Hungarian geography during the 1960s and 1970s. This was enabled by a combination of several factors, both domestic and international. Contributing to spatial planning and the efficient development of the (socialist) national economy has been a major goal for economic geographers since the Marxist-Leninist reorientation of the discipline after World War II, and many of these "new" geographers gained sufficient training to apply mathematical and statistical methods. Some scholars still influenced by interwar "bourgeois" geographers during their training also developed an interest in quantitative methods, which had not been entirely unknown in pre-Communist Hungary and which could reasonably be used in combination with a diverse set of methods from different geographical traditions. Foreign literature, guest scholarships, and conference attendances – both "Western" and "Eastern" – also provided access to mainstream quantitative works and inspired many scholars to regard quantification as fundamental to "modern science."

Those delivering quantitative spatial analyses in Hungary during the 1960s and 1970s formed a diverse group being trained at different university departments and working in different academic and economic planning institutions. They also differed in that some of them considered the quantitative approach as "the" future of spatial research, while to the eyes of others, it was but one of many traditions and toolkits in the long history of geography. The authors of quantitative works in those years were predominantly men, which was, however, not a distinctive feature of the quantitative approach but a general characteristic of Hungarian geography at that time.[12] Quantitative studies basically came from young scholars, but they still did not belong to the same generation. Whereas some key figures were born around 1930, others were born nearer to 1940 or even 1950. Some of the later were disciples of the first wave, but some were not. Very importantly, quantitative geography and regional science did not have that rapid decline in Hungary, which radical narratives suggest for the US and the UK. Instead, their role in scholarly research, as well as university curricula, kept gradually increasing, even in the 2000s, when Hungary's EU accession (2004) led to a boom in the need for geographers with quantitative skills in regional policy and planning, with postpositivist approaches starting to gain a stronger position only after the late 2000s.

Acknowledgment

The research has been supported by the János Bolyai Research Scholarship of the Hungarian Academy of Sciences.

Notes

1 In some domains of physical geography, especially those linked to astronomy, meteorology, and climatology, the use of sophisticated mathematical analytical methods was quite widespread. The circle of scholars focusing on these fields, however, was already rather separated from those working on social topics.
2 In the current chapter, all translations from Hungarian are mine.
3 Márton did not refer to Christaller, who otherwise was not unknown to contemporary Hungarian scholars (see, e.g., Prinz and Teleki 1937).
4 Kóródi was the youngest scholar in Hungary to ever have received the Doctor of Sciences (DSc) degree in the field of geography in 1969, when he only was 39. DSc is the highest scientific degree in the Hungarian academic system, issued by the Hungarian Academy of Sciences (see Péteri 1998).
5 For example, Klinghammer and Györffy (1973).
6 As Beluszky (born in 1936) explained in his recollections, when he started to use factor and cluster analysis in typologizing rural settlements, he asked a mathematician outside of his institute to help with calculations (Beluszky 2015).
7 Saushkin was the advisor of the US geographer Roland J. Fuchs during his stay at Moscow State University in 1960–1961 under a grant from the Inter-University Committee on Travel Grants (Fuchs 1964).
8 For data on concrete works, see Gyuris (2014b, 197).
9 Some seminal works on spatial analysis from foreign authors were also translated into Hungarian. These mainly included books from Soviet authors, but Isard's "Methods of Regional Analysis" was also translated (as "Methods of Regional Planning") in the National Planning Bureau's Institute of Economic Planning in 1973 (Isard 1973).
10 Enyedi worked in the Geographical Research Centre/Institute (GRI) of HAS from 1960 up until 1983. He then became chief director (1984–1992) of the newly established

multidisciplinary Centre for Regional Studies (CRS) of HAS, which had several departments in various regions of Hungary, and in the creation of which Enyedi played a leading role. Since 1984, CRS HAS and GRI HAS have coexisted as separate institutions, both employing several geographers.

11 Rechnitzer (2009, 13) writes about regional science as "a new branch of social science."

12 The share of women among the newly awarded Candidates of Sciences (CScs) in Hungarian geography, a lower scientific degree than DSc, was 0.0% (!) in the 1950s, 12.8% in the 1960s, and 14.3% in the 1970s, with a total of 111 new CScs (of both genders) between 1952 and 1980. (Own calculations based on data from volumes of the *Akadémiai Almanach*, the Almanac of the Hungarian Academy of Sciences.)

References

Abella, M. (1956): Vita a földrajzi tudományok filozófiai problémáiról. In *Földrajzi Értesítő*, 7(1–4), 462–466.

Ablonczy, B. (2007): *Pál Teleki: The Life of a Controversial Hungarian Politician*. Wayne, NJ: Hungarian Studies Publications.

Antal, Z. (2005): Metszetek a területfejlesztési szakemberképzés hazai történetéből. In Tatai, Z. (Ed.): *A területfejlesztés egyetemi oktatásáról*. Budapest: Területfejlesztő Öregdiákok Közössége, 135–139.

Antal, Z.; Perczel, G. (2005): Szemelvények tanszékünk történetéből. In Perczel, G.; Szabó, S. (Eds.): *100 éve született Mendöl Tibor*. Budapest: Trefort, 13–53.

Beluszky, P. (2015): "Mindig egy kicsit kívülálló voltam." Bögre Zsuzsanna és Kovács Teréz életút-interjúja Beluszky Pállal. In *SOCIO.HU*, 2015(1), 301–328; (4), 194–213.

Bernát, T.; Enyedi, G. (1961): *A magyar mezőgazdaság termelési körzetei*. Budapest: Mezőgazdasági Kiadó.

Boyce, D. (2004): A Short History of the Field of Regional Science. In *Papers in Regional Science*, 83(1), 31–57.

Czirfusz, M. (2015): Making the Space-Economy of Socialist Hungary: The Significance of the Division of Labor. In *Hungarian Cultural Studies*, 8, 105–123.

Dusek, T. (2010): Egy cikksorozat elé. Beszámoló "A Területi Statisztika fél évszázada" című műhelybeszélgetésről. In *Területi Statisztika*, 50(1), 3–5.

Dusek, T. (2014): Beszámoló a Kvantitatív forradalmak a területi kutatásban – egykor és ma, külföldön és idehaza című vitaülésről. In *Területi Statisztika*, 54(3), 300–302.

Enyedi, G. (1967): The Changing Face of Agriculture in Eastern Europe. In *Geographical Review*, 57(3), 358–372.

Enyedi, G. (1968): Területi különbségek a közép–kelet-európai agrárfejlődésben. In *Földrajzi Közlemények*, 92(2), 121–128.

Enyedi, G. (1973): A mezőgazdasági földrajz a 22. Nemzetközi Földrajzi Kongresszuson. In *Földrajzi Közlemények*, 97(3–4), 314–322.

Enyedi, G. (1976): Markos György halálára. In *Földrajzi Közlemények*, 100(1–2), 126–127.

Enyedi, G. (1984): Előszó. In Tamás Sikos, T. (Ed.): *Matematikai és statisztikai módszerek alkalmazási lehetőségei a területi kutatásokban*. Budapest: Akadémiai Kiadó, 9–10.

Fodor, F. (1925): A statisztikai ertékelés a gazdasági földrajzban. In *Földrajzi Közlemények*, 53(9–10), 202–210.

Fodor, F. (2006): *A magyar földrajztudomány története*. Budapest: MTA FKI.

Földrajzi Közlemények (1974): Beszámoló az Amerikai Egyesült Államokban, illetve a Szovjetunióban tett tanulmányutak néhány tapasztalatáról. In *Földrajzi Közlemények*, 98(2), 173–178.

Fuchs, R. J. (1964): Soviet Urban Geography: An Appraisal of Postwar Research. In *Annals of the Association of American Geographers*, 54(2), 276–289.

Geraszimov, I. P.; Preobrazsenszkij, V. Sz.; Szdaszjuk, G. V. (1977): A XXIII. Nemzetközi Földrajzi Kongresszus tudományos tevékenysége. In *Földrajzi Közlemények*, 101(4), 283–300.

Gerovitch, S. (2001): "Mathematical Machines" of the Cold War: Soviet Computing, American Cybernetics and Ideological Disputes in the Early 1950s. In *Social Studies of Science*, 31(2), 253–287.

Gyimesi, Z. (2014): The Contested Post-Socialist Rehabilitation of the Past: Dual Narratives in the Republishing of Tibor Mendöl's "Introduction to Geography." In *Hungarian Cultural Studies*, 7, 242–273.

Győri, R. (2009): Tibor Mendöl 1905–1966. In Lorimer, H.; Withers, C. W. J. (Eds.): *Geographers: Biobibliographical Studies 28*. London: Continuum, 39–54.

Győri, R.; Gyuris, F. (2012): The Sovietisation of Hungarian Geography, 1945–1960. In *Mitteilungen der Österreichischen Geographischen Gesellschaft*, 154, 107–128.

Győri, R.; Gyuris, F. (2015): Knowledge and Power in Sovietized Hungarian Geography. In Meusburger, P.; Gregory, D.; Suarsana, L. (Eds.): *Geographies of Knowledge and Power*. Dordrecht: Springer, 203–233.

Győri, R.; Withers, C. W. J. (2019): Trianon and Its Aftermath: British Geography and the "Dismemberment" of Hungary, c.1915-c.1922. In *Scottish Geographical Journal*, 135(1–2), 68–97.

Gyuris, F. (2009): A Teleki-iskola módszertani ujításai a két világháború közti magyar társadalomföldrajzban. In Sepsi, E.; Tóth, K. (Eds.): *Tudós tanárok az Eötvös Collegiumban*. Budapest: Ráció, 138–164.

Gyuris, F. (2014a): Human Geography, Cartography, and Statistics: A Toolkit for Geopolitical Goals in Hungary Until World War II. In *Hungarian Cultural Studies*, 7, 214–241.

Gyuris, F. (2014b): *The Political Discourse of Spatial Disparities: Geographical Inequalities between Science and Propaganda*. Cham: Springer.

Gyuris, F. (2018): Problem or Solution? Academic Internationalisation in Contemporary Human Geographies in East Central Europe. In *Geographische Zeitschrift*, 106(1), 38–49.

Gyuris, F. (2019): Az európai műveltség gyűrűi: egy revíziós térkép és eszak-amerikai előzményei. In Timár, E. L.; Berta, E.; Lehoczki, Z.; Pravetz, B. (Eds.): *KET. Kultúrák és etnikumok találkozása*. Budapest: Martin Opitz, 9–23.

Huntington, E. (1915): *Civilization and Climate*. New Haven, NJ: Yale University Press.

Huntington, E. (1927): *The Human Habitat*. New York, NY: D. Van Nostrand.

Illés, I. (1975): *Regionális gazdaságtan*. Budapest: Tankönyvkiadó.

Isard, W. (1973): *A regionális tervezés módszerei* (Translation). Budapest: Országos Tervhivatal Tervgazdasági Intézet.

Isard, W. (2003): *History of Regional Science and the Regional Science Association International: The Beginnings and Early History*. Berlin and Heidelberg: Springer.

Jefferson, M. (1911): The Culture of the Nations. In *Bulletin of the American Geographical Society*, 43(4), 241–265.

Jensen, R. G.; Karaska, G. J. (1969a): The Mathematical Thrust in Soviet Economic Geography: Its Nature and Significance. In *Journal of Regional Science*, 9(1), 141–152.

Jensen, R. G.; Karaska, G. J. (1969b): Application of Mathematical Methodology in Soviet Economic Geography. In *Soviet Geography*, 10(9), 501–506.

Jobbitt, S.; Győri, R. (2016): Ferenc Fodor: A Hungarian Geographer in the First Half of the Twentieth Century. In Győri, R.; Jobbitt, S. (Eds.): *Fodor Ferenc önéletírásai*. Budapest: ELTE Eötvös József Collegium, 39–77.

Jobbitt, S.; Győri, R. (forthcoming): Refugee Geographers: Gyula Prinz, Jenő Cholnoky, Ferenc Fodor and the Reimagining of Hungary after Trianon, 1920–1945. In Jobbitt, S.; Győri, R. (Eds.): *Geography and the Nation after Trianon*. London: Routledge.

Kádár, L. (1966): Elnöki megnyitó az MFT 90. közgyűlésén, 1966. május 12-én. In *Földrajzi Közlemények*, 90(3), 185–190.

Kádas, S. (1976): *A regionális modellezés irodalma*. Budapest: KSH Könyvtár és Dokumentációs Szolgálat.

Klinghammer, I.; Györffy, J. (1973): *Matematikai módszerek térképészek számára*. Budapest: ELTE Térképtudományi Tanszék.

Kóródi, J. (1959): *A borsodi iparvidék*. Budapest: KJK.

Kóródi, J.; Kovács, C. (1966): *A területi tervezés és irányítás az új gazdasági irányítási rendszerben (vitaanyag)*. Budapest. Manuscript.

Kőszegi, L. (1967): A gazdaság térbeli tervezése az új mechanizmusban. In *Földrajzi Közlemények*, 91(1), 25–44.

Kovács, C. (1962): Johann Heinrich von Thünen agrárföldrajzi jelentősége. In *Földrajzi Közlemények*, 86(1), 17–43.

Kovács, C. (1964): Árutermelő körzetek keletkezése a tőkés mezőgazdaságban heterogén természeti földrajzi alapokon. In *Földrajzi Értesítő*, 13(3), 315–339.

Kovács, C. (1966): A naturális és az arutermelő mezőgazdaság térbeli elrendeződésének főbb tényezői és törvényszerűségei. In *Földrajzi Értesítő*, 15(2), 211–235.

Kulcsár, V. (Ed.) (1976): *A regionális elemzések módszerei*. Budapest: Akadémiai Kiadó.

Lackó, L.; Erdős, G. (1970): Kartogramkészítés elektronikus számítógéppel. In *Geodézia és Kartográfia*, 22(5), 336–342.

Lengyel, I.; Nemes Nagy, J.; Rechnitzer, J.; Varga, A. (2020): A hazai regionális tudományról: eredmények és kihívások. In *Tér és Társadalom*, 34(1), 5–18.

Lux, G. (2015): *A területi modellezés története Magyarországon 1945 és 1990 között*. Budapest: MTA KRTK RKI.

Major, J. (1944): *Sopron város földrajza*. Doctoral Thesis, Manuscript. Budapest: Magyar Királyi Pázmány Péter Tudományegyetem.

Major, J. (1964): A magyar városhálózatról. In *Településtudományi Közlemények*, 13(16), 32–65.

Markos, G. (1951): A területi tervezés gazdaságföldrajzi megalapozása. In *Magyar-Szovjet Közgazdasági Szemle*, 5(4–5), 386–406.

Markos, G. (1955): Reflexiók egy beszámolóhoz. (A földrajzi tudományok rendszertani alapjairól). In *Földrajzi Közlemények*, 79(4), 359–365.

Markos, G. (1956a): Mérték és arány. In *Irodalmi Újság*, 5 May.

Markos, G. (1956b): Ember tervez – ember végez (széljegyzetek a tervvita margójára). In *Irodalmi Újság*, 2 June.

Márton, B. (1941): A kereskedő helységek eloszlása Magyarországon. In *Földrajzi Közlemények*, 69(2), 94–103.

Mathieson, R. S. (1969): The Soviet Contribution to Regional Science: A Review Article. In *Journal of Regional Science*, 9(1), 125–140.

Mendöl, T. (1936): A helyzeti energiák és egyéb tényezők szerepe városaink valódi nagyságában és jellegében. In *Földrajzi Közlemények*, 64(6–7), 98–108; (8–10), 121–132.

Mendöl, T. (1957): *Általános településföldrajz*. Budapest: Felsőoktatási Jegyzetellátó Vállalat.

Mészáros, J. (1975): Az I. Magyar – Amerikai Földrajzi Szeminárium. Budapest, 1975. május 26 – június 5. In *Földrajzi Közlemények*, 99(3–4), 355–358.

Miklós, G. (1972): Beszámoló a Magyar Földrajzi Társaság centenáriumáról. In *Földrajzi Közlemények*, 96(2–3), 234–237.

Móricz, F.; Abonyiné Palotás, J. (1975): *Matematikai módszerek a földrajzban*. Budapest: Tankönyvkiadó.

Nemes Nagy, J. (1973): *A gazdasági fejlettség területi különbségeinek elemzése matematikai és statisztikai módszerekkel*. Budapest: ELTE TTK.

Nemes Nagy, J. (1974): Changes of Regional Differences in the Economic Development Level in the Course of Socio-Economic Improvement. In *Annales Universitatis Scientiarum Budapestinensis de Rolando Eötvös Nominatae – Sectio Geographica*, 9, 61–84.

Nemes Nagy, J. (Ed.) (1976): *Regionális gazdaságföldrajzi olvasókönyv II. Az európai szocialista országok és a Szovjetunió gazdaságföldrajzi problémái*. Budapest: Tankönyvkiadó.

Nemes Nagy, J. (Ed.) (1977): *Regionális gazdaságföldrajzi gyakorlatok*. Budapest: Tankönyvkiadó.

Nemes Nagy, J. (2002): Területi elemzési ismeretek a geográfusképzésben. In Nemes Nagy, J. (Ed.): *Regionális Tudományi Tanulmányok 7: A Regionális Földrajzi Tanszék jubileuma*. Budapest: ELTE Regionális Földrajzi Tanszék, 49–54.

Németh, N. (2010): Múltidézés: a Területi Statisztika a hetvenes évtizedben. In *Területi Statisztika*, 50(1), 235–254.

Ormosy, V. (2013): Emlékezés Lackó Lászlóra (1933–1997). In *Tér és Társadalom*, 27(3), 172–177.

Pécsi, M.; Simon, L.; Szabó, P. Z. (1962): Az 1962. évi jubileumi földrajzi konferencia. In *Földrajzi Közlemények*, 86(4), 355–362.

Perczel, K. (1962): *A magyarországi regionális tervezés és a településhálózat fejlesztése*. CSc Thesis. Budapest.

Péteri, G. (1998): *Academia and State Socialism: Essays on the Political History of Academic Life in Post-1945 Hungary and Eastern Europe*. Highland Lakes, NJ: Atlantic Research and Publications.

Pokshishevskiy, V. V.; Mints, A. A.; Konstantinov, O. A. (1971): On New Directions in the Development of Soviet Economic Geography. In *Soviet Geography*, 12(7), 403–416.

Prinz, G. (1933): A földrajz az államigazgatás szolgálatában. In *Földrajzi Közlemények*, 61(4–6), 69–81.

Prinz, G.; Teleki, P. (1937): *Magyar föld, magyar faj 2. Magyar földrajz: A magyar munka földrajza*. Budapest: Királyi Magyar Egyetemi Nyomda.

Probáld, F. (1973): Területi különbségek Budapest éghajlatában. In *Földrajzi Közlemények*, 21(3–4), 229–251.

Probáld, F. (1974): A Study of Residential Segregation in Budapest. In *Annales Universitatis Scientiarum Budapestinensis de Rolando Eötvös Nominatae – Sectio Geographica*, 9, 103–112.

Probáld, F. (2002): A Regionális Földrajzi Tanszék 50 éve. In Nagy, J. N. (Ed.): *Regionális Tudományi Tanulmányok 7. A Regionális Földrajzi Tanszék jubileuma*. Budapest: ELTE Regionális Földrajzi Tanszék, 9–25.

Probáld, F. (2011): *Szakmai pályafutásom: emlékek a kezdetekről és a válaszutakról. (Vázlatos önéletírás a régmúlt 20. századból)*. Unpublished manuscript. Budapest.

Radó, S. (1962): A kommunizmus építése és a földrajzi tudományok. In *Földrajzi Közlemények*, 86(3), 225–232.

Rechnitzer, J. (2009): Bevezetés – a társadalomtudomány új ága, a regionális tudomány. In Lengyel, I.; Rechnitzer, J. (Eds.): *A regionális tudomány két évtizede Magyarországon*. Budapest: Akadémiai Kiadó, 13–24.

Rónai, A. (1989): *Térképezett történelem*. Budapest: Magvető.

Saushkin, Y. G. (1971): Results and Prospects of the Use of Mathematical Methods in Economic Geography. In *Soviet Geography*, 12(7), 416–427.

Sikos, T. T. (Ed.) (1984): *Matematikai és statisztikai módszerek alkalmazási lehetőségei a területi kutatásokban*. Budapest: Akadémiai Kiadó.

Simon, L. (1966): Főtitkári beszámoló. In *Földrajzi Közlemények*, 90(3), 265–270.
Tatai, Z. (1984): Markos György, a szocialista tervgazdaság szervezője és propagandistája. In *Egyetemi Szemle*, 6(1), 131–135.
Tatai, Z. (2004): Markos György. In *Comitatus*, 14(9), 71–80.
Teleki, P. (1917): *A földrajzi gondolat története*. Budapest: Own Publication of the Author.
Teleki, P. (1934): *Európáról és Magyarországról*. Budapest: Athenaeum.
Timár, J. (2020): Communist and Postcommunist Geographies. In Kobayashi, A. L. (Ed.): *International Encyclopedia of Human Geography*. Elsevier: Amsterdam, 2nd edition, 335–341.
Zaytsev, I. F. (1971): A Classification of Models in Economic Geography. In *Soviet Geography*, 12(2), 83–90.

7 A social history of quantitative geography in France from the 1970s to the 1990s

An overview of the blossoming of a multifaceted tradition

Olivier Orain

There is a massive gap between what the French "géographie théorique et quantitative" (TQG) means in Francophone countries and its blurry shape, provincial aura, and little notoriety abroad. It is of course very unspecific, as French, like German geography, stopped meaning anything in globalized geography for many decades. "French Theory" is by far more suggestive. However, in French context, "theoretical-quantitative" geography or "spatial analysis" – as it is alternatively called – played a role quite similar to its American counterpart and remain strong and influential 50 years after its rise during the 1970s. The main goal of this presentation is to provide an overview on the development of the French version of a more global movement, underpinning its peculiarities and its conformities. More specifically, this is an interesting example of what Jean-Marc Besse (2016) calls a "phase discrepancy" between two contemporary fields separated by language and, in some ways, cultural differences. Far from interpreting such a discrepancy in terms of lateness, my aim is to show how French geographers appropriated 15 years of anglophone debates and elaborations, mixed them with singularities from their inherited tradition and constructed a rather singular "quantitative" geography. At this point, the "quantitative" adjective attribute should be specified. Its mundane use isn't sufficient, because the recourse to and the measurement of quantities were already a systematic feature of geography. So I favor a sense that focuses on a distinctive act encompassing a methodology or a style of reasoning where the recourse to statistics or any other mathematical operations is the main way to show or explain something. And it obviously requires looking for the labels used to emphasize such a distinctive approach.

Indeed, it is slightly difficult to isolate a specific label that one could regard as the most generic and socially significant. Various players in the field used many different labels with different meanings, like "new geography," "theoretical and quantitative" geography, "spatial analysis," etc. (Cuyala 2014, 176–187). On the French scene, "quantitative geography" had been the initial label that its first promoters and opponents used to qualify such a promise (or threat), but it quickly shifted to "theoretical and quantitative," as many early promoters

DOI: 10.4324/9781003122104-7

emphasized the intrinsic "theoretical" demand of such an approach, so this contribution is entitled accordingly.[1] I'll consider all of them as congruent landmarks of a social movement that wished to initiate a radical change in geographical methodology (or epistemology), aiming to induce a shift based on explanation or the search for general laws, models, or patterns. Thomas Kuhn's theory of paradigms and scientific revolutions is often mobilized to summarize and reinterpret such a view (Orain 2009).

The literature on this topic is scarce, mixing actors' early contributions (e.g., Brunet 1976; Claval 1977; Pumain et al. 1983) or more recent ones (Claval 1998; Pumain and Robic 2002; Cauvin 2007), a PhD (Cuyala 2014), and the author's own work (Orain 2001, 2009, 2015, 2016). The choice of a social and epistemological approach is deliberate, first because this movement was mostly supported by groups imbued with a collective spirit partly inherited from the events of May–June 1968; and, second, in order to avoid an intellectualist rendition adding ideas and theories one after the other, out of social and cultural contexts. Basically, the chronology and the characteristics of such a movement are incomprehensible when deprived of their historical backgrounds. First, I will explain why a "scientific revolution" occurred in the 1970s, when different trends became explosive. Then I will analyze these "revolutionary" 1970s, keeping in mind that the actors remained a minority and that nothing like a replacement of the Vidalian paradigm occurred. Lastly, I will explore the climax (1981–1996) when "spatial analysis" seemed to take a leadership over the whole of French geography.

The conditions of a dramatic divide

Until 1968, French geography remained socially and intellectually cohesive.[2] The postwar decades had showed a dramatic increase of student and scholar numbers, the former multiplying by seven between 1949 and 1967 to respond to teaching needs (Soulié 2013). After the creation of a specialized university geographical curriculum during the Second World War, recruiting progressively shifted. Both academic and student bodies evolved toward a humble socioeconomic profile. Meanwhile, this population had followed high school scientific curricula more than the average in human sciences. The relationships between scholars and their pupils were less formal and closer than in traditional humanities, often metaphorized as those of a family. But, as in a family, there was also a paternalistic and authoritative dimension in these relationships. The student projects were closely bordered and watched by an academic community that had rigid, if not always self-justified, values. "This is not geography" was the main sentence that castigated attempts deemed too theoretical, speculative, and far away from "realities" (Robic et al. 2006; Orain 2009, 2015).

At the top of the disciplinary system was the prestigious Sorbonne, but geography was a more decentralized discipline than any other in humanity faculties and many provincial professors shone almost as much, thanks to a regional or national influence. The national so-called "school" had its prestigious journal, *Annales de Géographie*, and its national boards, every group gathering almost the same people. One generation, born in the beginning of the 20th century, reigned from the postwar period to the end of the 1960s. They were mostly men, politically (more or less) progressive; many were communists or Gaullists. Epistemologically, shortly, they reproduced unchanged and unquestioned old formulas, but allowed for a large expansion of new themes and objects, as far as it remained within due orthodoxy. Geomorphology remained the varsity team where one could prove their scientific skills, and also the main subfield for scientific innovations and controversies (Orain 2015; Bataillon 2020).

If scrutinized in details, this picture would show nuances, like outcast figures and temporary attempts to change the way geography was taught, practiced, or thought. But the main tendency was to thematically expand the discipline, not to reform it. Nevertheless, the rise of state planning during the 1960s and the subsequent rise of a transdisciplinary scene related to it confronted geographers with a full range of specialists they weren't used to speaking with: engineers, planners, economists, sociologists, etc. (Orain 2009, chap. 4–5). Among those professions, the rationale was very different: statistics, mathematical reasoning, and search for laws, were commonplace. Among the French elites involved in the management of state policies, the latter was ideologically and operationally dominant (Bezes et al. 2005). Out of a few personalities eager to reform geography to fit the new scientific climate, the main mood among the profession was to castigate such a "technocratic" view and to promote, in their terms, the wisdom of the one and only science that is "synthetic" first. The disciplinary literature, even in its reformist mode, is filled with numerous reassessments about the unique nature of geography as a crossroad of many scientific fields (Orain 2009, chap. 4–5). So it isn't surprising to see that French geography avoided the contemporary "quantitative" movement at its climax. Benefiting the growing wave of students and being able to offer them new jobs and opportunities, the scholars had an optimistic view on their discipline. They didn't notice growing signs of puzzlement and reluctance among Student Unions (Orain 2015), especially about the way pupils were used for applied geography surveys and "contracts" with no legal ground, in particular where it was broadly developed, like Strasbourg, Rennes, and Paris universities (Gaudin 2015; Usselman 2020; Orain 2015).

Nonetheless, "modernism" was a key attitude in this era, and statistics being a noticeable part of it, they started to be taught or used in the 1950s and 1960s, but sporadically, under some physical geographers' impetus, such as the climatologist Charles-Pierre Péguy (1915–2005). Generally, the main difficulty was that numbers were considered as values *per se* and symbols of magnitude in a description or a basic comparison. Pierre George (1909–2006) wrote a milestone article about the "myths of numbers" (1962) that is at the same time a very stimulating critique of such an attitude and the most consistent expression of a geographical reluctance to introduce mathematicized reasoning in the discipline. In the late 1960s, counting was everywhere as a gesture that gave a significant result out of a collective or individual survey, but the raw results seemed sufficient as they were a basis for charts and thematic cartography. They were scarcely a basis for the explanation of something, as explanations were local and seldom the main discursive scheme operating through a text or a speech.

Monolithic in its scientific features, demographically growing but mildly discredited by other social scientists, the French geography encountered a major social crisis at the end of the 1960s, during the May–June 1968 Crisis (Damamme et al. 2008; Gobille 2008a), like almost every sector of French society. It typically (Bourdieu 1990) opposed students and junior faculty members to the professorial body. May–June 1968 protesters demanded a radical shift in French politics and society, but their disciplinary goals were reformist, unquestioning the cult of fieldwork and the axiology inherited from previous generations (Orain 2015). Nonetheless, the May–June 1968 events showed quickly rising tensions between individuals or groups, and many despised teachers encountered stigma, caricatures, and ruckus, in a moment of generalized criticism and ideological critique. Every latent social contradiction became visible and problematic in the crisis dynamics (Dobry 2009 [1986]; Gobille 2008b). It initiated a big divide in what was previously a rather cohesive community.

Many scholars shifted their political views and became very hostile against the Left. On the other side, as a revolution, May–June 1968 failed to change French politics, even if the Gaullist power conceded socioeconomic improvements and a major reform of University system ("the Faure law," Poucet and Valence 2016). Following Pollak (1989) and Damamme et al. (2008), it is suggestive to interpret the changes that occurred in many parts of French society as the result of a reconversion of revolutionary initiatives at local and occupational scales, the protesters' sense of failure being conjured into more modest and proxemic initiatives. In that respect, French geography engaged in its particular "revolution." It involved either a new epistemological frame – the rise of a "theoretical-quantitative" movement seen as a way to replace an exhausted tradition – or a political critique against pointlessness, dull vacuity, and apolitical claims of the tradition.

The rise of a "revolutionary" movement (1970–1984): la "nouvelle géographie"

In the case of French geography, the "practical subversion" mood found a breach and a motive in the exhaustion of research formulas that the new generations had been bidden to use. A significant part of the newcomers in the field broke away from the community's tradition and social rituals. "New Geographers" were born at the end of the 1930s or belonged to the "baby boomers" (born between 1942–43 and 1950). They were numerous and socially diverging from the previous generations: coming from humbler backgrounds, having often followed "math'élem" (i.e. scientific) curriculum while in high school, as geography allowed a scholarship without Latin in the universities. They were more sensitive to "positivistic" values and had read the French philosophers of science like Bachelard, Canguilhem, and Althusser while in their senior year of high school. They were interested to and aware of statistics and mathematical reasoning. Before and after May–June 1968, they had been massively recruited as assistants in the growing universities. Later, they expressed retrospectively the lack of enthusiasm they felt while starting a doctorate in the 1960s and finding "old receipts" dull and not methodologically controlled (Chamussy 1978; Le Berre 1988).

In the years following May–June 1968, a formative offer in statistics, remote sensing analysis, and computational languages developed in Scientific institutions like Maison des sciences de l'homme, ORSTOM (Ultramarine Agency for Scientific and Technical Research), and CNRS (National Centre for Scientific Research). A significant minority of the newcomers showed a big interest for such training courses where elders didn't show up (Cuyala 2014). So they started early to socialize in a time where social stratification, enhanced by the recent political crisis, was sharp. In the French geographical landscape, it didn't develop in the universities where the seniors were still influential, charismatic enough, supposedly progressive, or very hostile. It produced a spatial differentiation of the reception of the new techniques: strong in East and Southeast of the country, where there was a relative lack of prominent professors; strong but in minority in Paris; influential in isolated and new university centers like Rouen and Reims; weak out of those social configurations.

Newcomers converted to such "new" techniques created affinity groups – like Groupe Dupont (1971), "Groupe d'action géographique" (GAG 1971–1972), and The Unicorn (mid-1970s) – related to the expansion of a new methodology. The latter one was a feminine group, which emphasized the significant role and demographic weight of women in the movement: as a non-mainstream, not socially valorizing, part of geography, quantitative techniques attracted a relatively strong inflow of young women (Pumain and Robic 2003; Ginsburger 2017). If mathematics didn't imply any political meaning *per se*, May–June 1968 projected shadow gave almost immediately a critical tone to such an initially reformist movement. Those groups were self-organized, autonomous from the mainstream and from the national discipline leaders. They offered an opportunity to remain "apart from the system" to the new generations and to operate what Boris Gobille calls a "symbolic dissidence" (Gobille 2008a) from the "catholic" society. For a decade, a political, epistemological, and social divide remained between the (scientific) revolutionaries and the geographical clergy, even if social relationships kept working episodically at a local level and with "modernist" leaders.

Meanwhile, the most reformist part of the senior body started to promote an acculturation of the French geography to the "novelties" coming from the USA and England. In the early 1970s, Philippe Pinchemel (1923–2008) initiated the translation of Brian Berry's *Geography of Central Markets and Retail Centers* (1971) and Peter Haggett's *Locational Analysis in Human Geography* (1973). The main way to promote a new methodology and to appropriate the Anglophone "model" remained the auctorial mediation through conferences, articles, and books. In the late 1960s, Paul Claval (b. 1932) developed a specialized anglophone library in Besançon University where he invited many junior geographers. Lecturers who had been working in Canada or the USA in the 1960s, like Bernard Marchand (b. 1934), Sylvie Rimbert (b. 1927), Henri Reymond (1930–2018), and Jean-Bernard Racine (b. 1940), there converted to the "quantitative" techniques, became the prominent publicists of the American "quantitative techniques." The former taught a "quantitative geography" certificate at Paris University in 1969 and gave a lecture in 1970 in a national disciplinary gathering, followed by many newcomers, who became legendary for the movement (Marchand 1972). Many actors later narrated how this speech bedazzled them and accelerated their conversion to the new tools and ways of thinking (Chamussy 1978; Le Berre 1988). Racine & Reymond wrote a handbook, *L'Analyse quantitative en géographie* (1973) that quickly became a milestone, even if the readers retained mainly the theoretical dimension of the book. S. Rimbert played a seminal role in the diffusion of quantitative methods in cartography by her collaborative work with Colette Cauvin (b. 1944). François Durand-Dastès (1931–2021) had been another early promoter, especially for modeling approaches. Famous or not, some Anglophone geographers have been invited in France through personal contacts, like Peter Haggett, Peter Gould, Wanda Herzog, and a few others.

In 1970, Roger Brunet (b. 1931), already a prominent modernist figure, author of a notorious state doctorate and a no-less notorious secondary thesis about discontinuity phenomena in geography (1965) started to work on the launching of a new scientific journal. He and other modernists became convinced that *Annals of Geography* weren't able to host such an *aggiornamento* (update). He wrote an editorial, read and slightly revised by his peers, and gathered material for a prototype. The first issue of *L'Espace géographique* came in 1972 as an academic journal, with a publisher specialized in medicine (Doin). The first year provided many articles about the then-called "crisis" of French geography, about quantitative initiation or methodology discussions, topics that were avoided by mainstream journals. Brunet was a reformist, with a socialist sensibility. The first editorial was more concerned with epistemological debates and self-reflexivity geographers should develop than by the "quantitative" aspect *per se*. The journal hosted in the following years many new developments and debates about Marxism, semiology, perceptual or phenomenological geography (the so-called "lived space" promoted by Armand Frémont [1933–2019]), physically integrated landscapes advocated by Toulousian geographer Georges Bertrand, etc. Nonetheless, *L'Espace géographique* quickly became the counter-mainstream journal and the voice of the then French "New Geography" in its diverse foci and tendencies. Many other groups developed during the 1970s with a more radical tone and various political critiques against French society or practiced geography. "New Geography" had been a very diverse movement.

The political climate in the community deteriorated quickly. Jean Bastié (1919–2018), a former Communist and Pierre George's student, traumatized by May–June 1968, rallied all the most conservative scholars into the Syndicat autonome (Autonomous Trade-Union) and initiated a witch hunt against anything "leftist" in the discipline, stopping careers, rebuking grants to any "suspect" project, etc. (Burgel et al. 1986). The climate became tenser than it already was. In 1972, the liberal Prime Minister Jacques Chaban-Delmas was replaced by Pierre Messmer, a more authoritarian and conservative figure. He stopped the "New Society" project promoted by his predecessor that tried to respond to social attempts expressed in May–June 1968 and initiated a drastic retraction of public credits in the University system. After 1973, the oil shock and pending economic crisis aggravated the general situation. So the climate became heavy and criticism developed among new geographers and their opponents. In 1975, a group of students at the École normale supérieure de l'enseignement technique (ENSET) in Cachan launched a fanzine called *EspacesTemps*, devoted to a radical critique of history and geography. Mixing an ideological and an epistemological critique inspired by Althusser, the mimeographed pamphlet had a radical tone. A few other radical fanzines showed up in the following years, like *Attila* (1976) and *Espaces & Luttes* (social struggles) (1978), but they remained ephemeral (Orain 2009, 2015).

In 1976, Yves Lacoste (b. 1929) launched *Hérodote*, another critical journal published with the more substantial support of left-wing librarian and publisher François Maspéro (Hepple 2000). He was already a notorious geographer, author of famous books on underdevelopment geography and senior lecturer at countercultural university of Vincennes (Soulié 2012). In 1976, Lacoste also wrote a lampoon whose main argument was political, denouncing academic geography as a boring veil occulting the true one, practiced by armies and their general staffs, with a hint of conspiracy theory (Lacoste 1976). The same year, the Groupe Dupont and a few "Swiss" professors organized the first symposium called Géopoint in Geneva, hosted by Claude Raffestin (born in 1936) and J. B. Racine. It gathered more than 100 "new" geographers from France, Belgium, and Switzerland with a very "May–June 1968" functioning style, with an equalization of every attendant and a passionate debating atmosphere (Orain 2009, chap. 6). "Quantitative" geography was the main positive reference and its new practitioners and converts were a majority among the attendance. The Géopoint symposia continued every two years with their various mix of "positivistic" and "Marxist" aims as already theorized by H. Reymond in his 1976 contribution (Reymond 1976). It became a perennial gathering until 2018 with significant evolutions from the 1980s.

The years 1976–1978 remain as the climax of a polemic time, between Mandarins (like P. George, M. Le Lannou [1906–1992], or Numa Broc [1934–2017]) and Young Turks (like Jacques Lévy [b. 1951]) or between leftist groups (Robic 1998). In an attempt to provide a more consensual manifesto, four Groupe Dupont members wrote, "Espace, que de brouillons comment-on en ton nom!" (Chamussy et al. 1977) in the first issue of another fanzine called *Les Brouillons Dupont*, in 1977, but it remained unseen and by large ignored by mainstream geography at the time. This text reflects with sound clarity and a quieted tone where the "theoretical-quantitative" movement was conceptually and epistemologically in the late 1970s.

New geography produced schematically three main critiques against what had done the "old" "traditional" one (Orain 2009, chap. 6). They were modulated and diversely combined by the various groups and individuals that endorsed the "scientific revolution" during those critical years. They shared the idea that geography was a political inquiry, even if traditional scholars claimed the neutrality of it. They advocated the clarification of implicit ideological motives and political aim. They called for a discipline useful for people. The second point was about realism and constructs: following Raffestin's seminal conference in Géopoint 76 (Raffestin, 1976). Reader of a wide culture, familiar with epistemology, semiology, and social sciences, he provided then a preliminary critique of geographical realism with nominalist and linguistic arguments inspired by Greimas and Prieto. He displayed the idea that geography scrutinizes reality through language and "constructs," nothing being done *per se*. His ideas on such a question had got a large response during the two first Géopoint symposia. But the most important point for the "quantitative" attendance was that the classical geography was "prescientific" (in Bachelardian acceptation) or wasn't scientific at all. Most of them shared a normative vision of science as the search for laws with a rigorous methodology and mathematized techniques.

Their views on the latter made reference to the Anglophone classics, like Haggett (1965) or Abler et al. (1971). But as practitioners, they learned techniques that were available in training or summer courses and self-teaching seminars (Cauvin 2007; Cuyala 2014). It favored an early emphasis on Multivariate statistics in the form it took in France under Jean-Paul Benzécri's reference and Claude Deniau's (b. 1940) supervision: Correspondence analysis and Hierarchical cluster analysis. Something like an alliance developed between quantitative geographers and mathematicians eager to offer mathematized techniques for social sciences,[3] under the Marc Barbut's (1928–2011) leadership at the École des Hautes Études en Sciences Sociales (EHESS) (see Cauvin 2007). The French "theoretical-quantitativists" progressively enlarged their skills in a broad array of techniques (e.g., topology, temporal statistics).

They were fascinated by General System Theory (GST) popularized in France by many vulgarizers (e.g., J. de Rosnay, J. L. Le Moigne, E. Morin, Y. Barel) and especially congruent with a tradition to see a "region" or an "environment" as a "combination" of various evolving factors (Orain 2001). During the 1970s, it was a general theoretical scheme rather than a basis for mathematical reasoning. One of the early doctoral dissertations in the new style defended in France, Franck Auriac's (1935–2018) *Système économique et espace. Un exemple en Languedoc*, provides a good illustration. This dense and ambitious work uses the GST in a nonquantitative, verbal, reconstruction of the Languedocian vineyard history with a Systemic and Marxist theoretical frame (Orain 2001). However, the quantitative tools of multivariate analysis are applied to various data to interpret inductively the "spatiality" of the vineyard, with no effective applying of any "systemic" proceedings or mathematical models. The only model, implicit in such a denomination, is the gradual center-periphery one, which Auriac uses graphically to interpret the spatial organization of the vineyard and its backward effect on spatial diffusion of innovation. At such a stage, modeling wasn't a discussed and pervasive practice, even if French theoretician-quantitativists were well aware of them (Durand-Dastès 1974), being readers of Chorley and Haggett (1966) and many other anglophone works using this range of methods. Denise Pumain's (b. 1946) doctorial dissertation, *Contribution to the Study of Urban Growth in the French Urban System* (Pumain 1980), is probably the first one to "test" one, the Gibrat's rule of proportionate growth.[4]

In its first years, the new movement was spontaneous, multifaceted, and stratified, moved by a genuine curiosity for anything new and foreign. TQG national gatherings were organized every two years in Besançon (Cuyala 2014), in turn with Géopoint symposia (more theoretically oriented). Grants allowed were tiny but sufficient, out of a bigger DGRST[5] – the General Board for Scientific and Technical Research, the main fund then in France – contract obtained in 1974. In 1978, Rimbert, who was already a senior and had many acquaintances in Europe and North America (Cuyala 2014, 2016), organized a tentative TQ European colloquium in Strasbourg that succeeded more than expected and initiated the prolongation of such gatherings every two years, becoming the European Colloquium on Theoretical and Quantitative Geography (ECTQG). In this early stage of development, the French TQG was an active but small minority and a dominated movement, far away from institutional power or wide funding, castigated by the establishment and seen as a dangerous leftist thing. The power balance changed with the François Mitterrand's presidential election in 1981 (Robic 2006). The movement was matured enough and empowered by its inner activism. The conjunction of political change and self-structuring dynamics opened a new era.

Quantitative geography rethought as "spatial analysis": a prosperous era (1984–1996)

After almost 25 years of right-wing governments in France, François Mitterrand won the 1981 presidential election, followed by a Landslide in legislative elections in June 1981. The Socialist scientific politics totally changed how research and universities were regulated and financed: in 1982 the ministry of research deeply reformed CNRS, making researchers civil servants with a real public status, dramatically increasing the financial support to research teams and projects, widening recruitments, reducing state-designated seats in the committees, changing most of the high administration managers. In 1984, the Savary law operated equally dramatic changes in the university system, creating a stable tenured junior status (literally "master of conferences"), terminating the state doctorate, and creating a new doctorial formula (closer to PhD), creating a global joint venture status for CNRS university teams, and reforming the National University Conference[6] in a more inclusive and democratic spirit. Academic autonomy was a pillar of those reforms, the financial effort was another. France knew at the time what is called the "second massification" of university system and socialist governments allowed the reopening of junior scholars' large recruitments and a significant expansion of senior tenures. At another level, socialist management promoted promethean research politics that called many new initiatives.

In this climate, political decisions eventually allowed funds and status to geographical teams that were snubbed previously. The "Groupe Dupont" became a "young team" in 1982, a status endorsed and financed by CNRS. Many other tendencies showed up like "social geography" or Lacoste's (from now on) "geopolitics" that could now apply for public recognition and grants. It was also related to a slow sociopolitical change: the first doctorial dissertations in TQG styles had been defended in the late 1970s, other progressive geographers became professors and the national boards elections progressively shifted, as the younger generations progressively took over the latter ones, relegating to a minority the "autonomous" (i.e., right-wing) lists. In the first years of socialist power, the political tensions in the geographical community intensified before a decrease around 1984 Paris IGU international congress. In those years, Brunet, among others, played a key-role in reversing the equilibrium in geographical community. Then close to Jean-Pierre Chevènement's movement (CERES) in Socialist party, he worked in successive cabinets in the Ministry for Research and Industry (1982–1984). The exact role Brunet played in the higher administration has still to be studied through archives and he wasn't the one and only geographer in an executive position. This administrative position somewhat underpinned a prolific period of publishing that started a bit before (in 1980) and continued until the mid-1990s, giving him his specific position in the history of French geography as a spatial analysis controversial leader.

In 1980, after five silent years, he published "La Composition des modèles en analyse spatiale" (Brunet 1993). This article systematized his early authorial proposals and hosted, articulated, and reconstructed many symbolic generalizations produced by spatial theoreticians, in a very idiosyncratic way (Orain 2016). Significantly, he avoided the "TQG" label mainly endorsed in the social movement – which he wasn't strictly a part of. He offered a unified theory of "spatial analysis" using linguistic analogy and implicit structuralist motives. The central idea was that spatial organization of any geographical object is a combination of "elementary structures of space" (Levi-Straussian nod), called "chorèmes," a neologism forged from the terminology of linguistics. Haggett's (1965) analytic interpretation of spatial human geometry (seedlings, networks, areas) and processes (interaction, diffusion, hierarchy) was discretely at the core of such a construction. In 1984, he left his function in Research ministry to build a flagship platform of projects in Montpellier, the Public Interest Group (GIP in French) RECLUS – an acronym for "Network for location and spatial units study." This was also a publishing house and an economic contractor. He then became the "industry captain" of French geography (Lévy 1996), producing a new *Universal Geography* (UG) collection (1989–1996), launching a cartographic journal, *Mappemonde* (1986) and atlas collections, a dictionary (1991), among other activities. This structure allowed him to fund many researches in TQG and to recruit scholars. In the 1980s and 1990s, *L'Espace géographique* appeared more and more as the "spatial analysis" tribune, even if this stereotype was incorrect and disregarding the variety of themes and tendencies published in the journal (Guérin-Pace and Saint-Julien 2012). More and more popular (or disliked) and notorious among French geographers, Brunet became in common knowledge the "pope" of TQG/spatial analysis even if it came from a slight misunderstanding. A significant part of TQG practitioners got involved in the *UG*'s 10 issues, even if many "classical" geographers had also been recruited. In the first UG volume, *New Worlds* (Brunet 1990), he published a general treatise "The Deciphering of the World" that is his contribution to a theory of space and geography.

Brunet isn't the only theoretician of space that arose in France from TQG: in the late 1970s, some other thinkers developed such an attempt. Georges Nicolas (b. 1932) searched for geographical axiomatization under the auspices of the Bourbaki collective and logical positivism (Nicolas 1984). Henri Reymond developed a less totalizing but intellectually challenging attempt that remained little noticed (Reymond 1981). Pinchemel developed a more accessible and comprehensive grand theory that used more common references and languages: first through articles, then a "handbook" published his retirement year, as a testimony and an *opus maximus*, *The Face of the Earth* (Pinchemel and Pinchemel 1988). In many ways, Brunet's and Pinchemel's elaborations are similar and congruent, coming from the same classical matrix and aiming at a similar unified, global, perspective. For both of them, mathematical reasoning and statistics were only rational tools, not a language.

While Brunet's and Pinchemel's French space theory was partly idiosyncratic and deeply rooted in local cultural inherited tradition, French "quantitativists" early tried to get involved in European gatherings and activities. Those events offered many opportunities to open on new techniques, especially in modeling, computing, and data treatments. Being involved in an international network led the French attendants to distance themselves from the previous interests they had developed and to partially normalize their research styles. In the 1980s, many members of Groupe Dupont got involved in an attempt to make their systemic approach a mathematized one, first with Forrester methods and computer programming. They produced the AMORAL model (pun intended), financed with regional funding and fiercely promoted, but with little results. In Pumain's and Durand-Dastès' trail, Lena Sanders (b. 1955) developed a dynamic modeling using differential equations (Sanders 1984). The passing from basic recurrence relations to differential calculus is an example of the growing sophistication TQG gained during the 1980s. City systems were a key topic in Pumain's school opening on a wide range of methodologies and techniques.

The movement as a whole endorsed many other techniques and invested areas like physical geography, network analysis, etc. The wide diffusion of computers in 1980s France helped to enhance the array and the speed of researches (Pumain et al. 1983). The conjunction of socialist largesse, foundation of dedicated laboratories, and specialized teaching made the movement able to stabilize in an institutionally recognized form. For a few years (1984–1993), it occupied a leading role in French geography and appeared as a wannabe new "normal" science. But its practitioners never became a majority: they remained under 10% of the geographical body,[7] even during the huge demographic growth the movement knew between 1984 and 1999, doubling its numbers (Cuyala 2014). This assessment needs a further and ultimate examination on the grounds and limits of TQG/spatial analysis ephemeral dominance.

Conclusive epilogue

In the 1980s–1990s, many new tendencies arose and the previous monolithic discipline eventually shifted into a diversified, piecemeal field structured through currents. Having started in the 1970s, TQG was the most advanced tendency in terms of structuring. It had visible leaders, a strong social network, institutional recognition, functional and ritual gatherings. In this period, laboratories and graduate teaching programs flourished in various places. From the end of the 1980s, it allowed TQG to produce new generations of practitioners, early trained and skilled, and socialized in a wide research network. At a graduate and postgraduate level, it gave an opportunity to define concrete problem solutions and to elaborate a progressive apprenticeship for students, donating in the same journey shared values, symbolic generalizations, metaphysics and, of course, techniques (Kuhn 1970).

Research styles were slightly different from place to place but there was a global homogeneity among French trainings. However, TQG was just a part of French geography, remaining in a heteronomous position, both locally and nationally. It recruited mostly bright but few students. Geography being a human science in French curricula, most of the students and teachers were repelled by mathematics and prone to avoid them. New currents like geopolitics, humanistic, social, environmental, or cultural geographies in some ways grew up against TQG that became the new *apparently* hegemonic tendency. Something like a backlash occurred in the end of the 1980s and produced a sharp divide in many local situations. The scientific values expressed by TQG prominent figures, their perceived "arrogance," brought up a high level of reluctance among their colleagues in various local situations. TQG scholars failed to explain to their colleagues and student basis what was useful and significant out of quantitative frames in spatial analysis, especially a space theory.

Globally, TQG actors were solid debaters and among the brightest minds in the disciplinary landscape but appeared unable to make their way socially and cognitively acceptable for a broader audience. When right-wing parties won the 1993 legislative elections, the backlash became materially ostensive: the new disciplinary executives obtained GIP RECLUS dismantlement. The cohesive nature of TQG movement, its international openness, and the institutional resiliency of French research structures protected the movement against the shrinking processes it encountered in many other countries. Some local schools, like the one gathered around Denise Pumain, Thérèse Saint-Julien (b. 1941), and Lena Sanders, became widely influent, scattering among French and even European universities. The former is probably one of the most famous TQG authors internationally, through alliances with complex systems research and archaeologists. The provincialized status of French geography operates as a limitation, in an international field that is itself provincialized and disvalued, and, in TQG case, competed by regional science, spatial economy, and even physics. Interdisciplinary projects became the way to bypass such a difficulty.

The attempt to provide a dialectic history, both social and cognitive, of what happened to be a minority but major movement in French context, shows how social and scientific aspects are deeply intertwined. I emphasized in turn what was peculiar or similar in such a case. French TQG has certainly been a contrasted odyssey that opened a pluralist and theoretically oriented era. They successfully operated an adjustment on international standards, the first current in France doing so. But, maybe, what could surprise a foreign reader and open a renewed interest maybe locates in what had been deeply and peculiarly French in it. The spatial theories, elaborated during the late 1970s and the 1980s, remain as a significant heritage of a creative period.

Notes

1 The history of the various uses of this label(s), their temporal and individual or social variations are of great interest, but I decided to consider them as almost synonymous to simplify the presentation.
2 For more details and analysis about the 1945–1968 period, see Orain (2015).
3 For the few mathematicians making a bet on applied techniques for social sciences, finding welcoming groups was of crucial value as their career was at stake in terms of professional axiology. In a very platonistic and selfless community that was at the top of symbolic academic hierarchy (Bourdieu 1984), working for such "low" disciplines with down-to-earth goals was downgrading. To compensate such a marginalizing effect, they had to enroll as many converts as they could, getting symbolic reassessment and "customers" for stages, books, and other financial retributions.
4 This law stipulates that in an economic system every unit P_i grows during a time period under a proportional factor, noted r, common to all the units, with a marginal variation. It is expressed as $dP_i = rP_i + \varepsilon$. After many time periods, various r factors applied to ε could increase the differences between the units. Gibrat's law is a way to explain the dynamics of differentiation in a population of cities, enterprises, or areas. In Pumain's state doctorate, Gibrat's law isn't "tested" in a Popperian fashion but verified through proportionality relations between cities within census data.
5 General direction for scientific and technical research, created in 1958, had been a major funding tool between 1958 and the 1980s.
6 This institution, divided in disciplinary half-elected, half-nominated, councils had, among other tasks, to certify one candidate's ability to become a professor. It increased its labelling process to senior lecturers (maîtres de conferences) in 1984.
7 In his PhD dissertation, S. Cuyala has calculated the balance of various affiliations through geographical directories published every five years and based on self-declarations (Cuyala 2014).

References

Abler, R.; Adams, J. S.; Gould, P. (1971): *Spatial Organization: The Geographer's View of the World*. Englewood Cliffs, NJ: Prentice Hall International Inc.

Auriac, F. (1979): *Système Economique et Espace. Un Exemple en Languedoc*. State doctorial dissertation. Montpellier: University of Montpellier. Published in 1982 as *Système Economique et Espace*. Paris: Economica, "Geographia", 3.

Bataillon, C. (2020): Le Système Institutionnel des Géographes de France et 1968. In Orain, O.; Robic, M.-C.: Mai-Juin 68, *L'Espace Géographique* et la Mémoire d'une Communauté. Special Issue. In *L'Espace Géographique*, 2020(1), 30–34.

Berry, B. J. L. (1967): *Geography of Market Centers and Retail Distribution*. Englewood Cliffs: Prentice Hall.

Berry, B. J. L. (1971): *Géographie des Marchés et du Commerce de Détail*. Paris: Armand Colin, French, translated by B. Marchand.

Besse, J. M. (2016): Founding Landscape Studies: John Brinckerhoff Jackson and French Human Geography. In *L'Espace Géographique*, 45(3), 195–210.

Bezes, P.; Chauvière, M.; Chevallier, J.; Montricher, N.; Ocqueteau, F. (2005): *L'État à L'épreuve des Sciences Sociales. La Fonction Recherche Dans les Administrations Sous la Vᵉ République*. Paris: La Découverte, "Recherches".

Bourdieu, P. (1990 [1984]): *Homo Academicus*. Cambridge: Polity Press.

Brunet, R. (1965): *Les Phénomènes de Discontinuité en Géographie*. Complementary dissertation of state doctorial dissertation. Toulouse: University of Toulouse. Published under the same title in 1967. Paris: CNRS, "Mémoires et documents".

Brunet, R. (1976): Rapport sur la "*New Geography*" en France. In Clark, J. I.; Pinchemel, P. (Eds.): *Geography in France and in Britain*. London: Institute of British Geographers, SSRC, 40–44.

Brunet, R. (1990): Le Déchiffrement du Monde. In Brunet, R.; Dolfuss, O. (Eds.): *Mondes Nouveaux. Géographie Universelle*, v.1. Part 1, Paris-Montpellier, Belin-Reclus, 1990. Reissued and revised in 2001 as *Le Déchiffrement du Monde. Théorie et Pratique de la Géographie*. Paris: Belin, Mappemonde.

Brunet, R. (1993): Building Models for Spatial Analysis (1980). In *L'Espace Géographique*, Special Issue, "Espaces, Modes d'emploi. Two Decades of l'Espace Géographique, an Anthology. Special issue in English", 109–123.

Burgel, G.; Rochefort, M.; Seronde-Babonaux, A. M. (1986): Témoignage d'une Renaissance. In Beaujeu-Garnier, J. (Ed.): *Sens et Non-Sens de L'espace: de la Géographie Urbaine à la Géographie Sociale*. Caen: Collectif français de géographie urbaine et sociale, 9–10.

Cauvin, C. (2007): Géographie et Mathématique Statistique, une Rencontre d'un Nouveau Genre. In *Revue Pour L'histoire du CNRS*, (18), 15–19.

Chamussy, H. (1978): D'amour et D'impuissance. In *Brouillons Dupont*, (3), 67–81.

Chamussy, H.; Charre, J.; Durand, M.-G.; Le Berre, M. (1977): Espace, Que de Brouillons Commet-on en ton Nom!. In *Brouillons Dupont*, (1), 15–30.

Chorley, R.; Haggett, P. (Eds.) (1966): *Models in Geography*. London: Methuen.

Claval, P. (1977): *La Nouvelle Géographie*. Paris: PUF, Que Sais-je?

Claval, P. (1998): *Histoire de la Géographie Française de 1870 à nos Jours*. Paris: Nathan Université, Réf.

Cuyala, S. (2014): *Analyse Spatio-Temporelle d'un Mouvement Scientifique. L'exemple de la Géographie Théorique et Quantitative Européenne Francophone*. PhD. Paris: University of Paris 1.

Cuyala, S. (2016): The Spatial Diffusion of Geography: A Bibliometric Analysis of ECTQG Conferences (1978–2013). In *Cybergeo. European Journal of Geography* [Online], Epistemology, History of Geography, Didactics, document 783, online 15 June, read 22 April 2020.

Damamme et al. (2008): *Mai Juin 68*. Paris: Les éditions de l'atelier.

Dobry, M. (2009 [1986]): *Sociologie des Crises Politiques. La Dynamique des Mobilisations Multisectorielles*. Paris: Presses de Sciences Po, Fait Politique, 3rd edition.

Durand-Dastès, F. (1974): Quelques Remarques sur L'utilisation des Modèles. In *Bulletin de L'Association de Géographes Français*, 51(1), n.413–414, 43–50.

Gaudin, S. (2015): Le Temps de L'engagement, Enjeux et Développement d'une Géographie Appliquée (1970–1980). In *Bulletin de L'association de Géographes Français*, 92(1), 111–125.

George, P. (1962): Quelques Aspects des Mythes du Nombre. In *Cahiers Internationaux de Sociologie*, 17(2), n.33, 39–47.

Ginsburger, N. (2017): Women in Geography in an Era of Change: Feminisation and Feminism in the French and International Academic Field (1960–1990). In *L'Espace Géographique*, 46(3), 236–263.

Gobille, B. (2008a): La Vocation d'hétérodoxie. In Damamme, D.; et al.: *op. cit.*, 274–291.

Gobille, B. (2008b): L'événement Mai 68. Pour une Sociohistoire du Temps Court. In *Annales. Histoire, Sciences Sociales*, 63(2), 321–349.

Guérin-Pace, F.; Saint-Julien, T. (2012): The Words of *L'Espace Géographique*: A Lexical Analysis of the Titles and Keywords from 1972 to 2010. In *L'Espace Géographique*, 41(1), 4–31.

Haggett, P. (1965): *Locational Analysis in Human Geography*. London: Arnold.

Haggett, P. (1973): *L'analyse Spatiale en Géographie Humaine*. Paris: Armand Colin, French, translated by Hubert Fréchou.

Hepple, L. W. (2000): Yves Lacoste, Hérodote and French Radical Geopolitics. In Atkinson, D.; Dodds, K. (Eds.): *Geopolitical Traditions: Critical Histories of a Century of Geopolitical Thought*. London: Routledge, "Critical Geographies", 268–301.

Kuhn, T. S. (1970): Postscript. In Kuhn, T. S. (Ed.): *The Structure of Scientific Revolutions*. Chicago: University of Chicago Press, 2nd edition.

Lacoste, Y. (1976): *La Géographie, ça Sert D'abord à Faire la Guerre*. Paris: François-Maspero, Petite Collection Maspero.

Le Berre, M. (1988): Itinéraire Géographique. Vingt ans Après. In *Brouillons Dupont*, (17).

Lévy, J. (1996): Roger Brunet. In Juillard, J.; Winock, M. (Eds.): *Dictionnaire des Intellectuels Français*. Paris: Le Seuil, 192–193.

Marchand, B. (1972): L'usage des Statistiques en Géographie. In *L'Espace Géographique*, 1(2), 79–100.

Nicolas, G. (1984): *L'Espace Originel. Axiomatisation de la Géographie*. Berne: Lang.

Orain, O. (2001): Démarches Systémiques et Géographie Humaine. In Robic, M.-C. (Ed.): *Déterminisme, Possibilisme, Approche Systémique: les Causalités en Géographie*. Vanves: CNED, Part 3, 1–64.

Orain, O. (2009): *De Plain-Pied Dans le Monde. Écriture et Réalisme dans la Géographie Française au* XXe *Siècle*. Paris: L'Harmattan, Histoire des sciences humaines.

Orain, O. (2015): Mai-68 et Ses Suites en Géographie Française. In *Revue D'histoire des Sciences Humaines*, (26), 209–242.

Orain, O. (2016): Le Rôle de la Graphique dans la Modélisation en Géographie. Contribution à une Histoire Épistémologique de la Modélisation des Spatialités Humaines. In Blanckaert, C.; Léon, J.; Samain, D. (Eds.): *Modélisations et Sciences Humaines. Figurer, Interpréter, Simuler*. Paris: L'Harmattan, Histoire des Sciences Humaines, 215–268.

Pinchemel, P.; Pinchemel, G. (1988): *La Face de la Terre*. Paris: Armand Colin.

Pollak, M. (1989): Signes de Crise, Signes de Changement. In *Cahiers de l'IHTP*, "Mai 68 et les Sciences Sociales", n.11, April, 9–20.

Poucet, B.; Valence, D. (Eds.) (2016): *La Loi Edgar Faure. Réformer L'université Après 1968*. Rennes: Presses Universitaires de Rennes, Histoire.

Pumain, D. (1980): *Contribution à L'étude de la Croissance Urbaine dans le Système Urbain Français*. State doctorial dissertation. Paris: University of Paris 1. Published as *La Dynamique des Villes*. Paris: Economica, 1982.

Pumain, D.; Robic, M.-C. (2003): Le Rôle des Mathématiques dans une "Révolution" Théorique et Quantitative: la Géographie Française Depuis les Années 1970. In *Revue D'histoire des Sciences Humaines*, (6), 123–144.

Pumain, D.; Saint-Julien, T.; Vigouroux, M. (1983): Jouer de L'ordinateur sur un Air Urbain. In *Annales de Géographie*, 92(511), 331–346.

Racine, J.-B.; Reymond, H. (1973): *L'analyse Quantitative en Géographie*. Paris: P.U.F., "Sup" Le Géographe, n.12.

Raffestin, C. (1976): Problématique et Explication en Géographie Humaine. In Groupe Dupont (Ed.): *Géopoint, Théories et Géographie*. Genève: Universités de Genève et Lausanne, 55–73.

Reymond, H. (1976): La Nécessité de Discuter d'un Contenu Explicite et Critique. In Groupe Dupont (Ed.): *Géopoint 76, Théories et Géographie*. Genève: Universités de Genève et Lausanne, 106–111.

Reymond, H. (1981): Une Problématique Théorique Pour la Géographie: Plaidoyer Pour une Chorotaxie Expérimentale. In Isnard, H.; Racine, J.-B.; Reymond, H. (Eds.): *Problématiques de la Géographie*. Paris: P.U.F., 163–249.

Robic, M.-C. (1998): Dupont et les Autres. In *Brouillons Dupont*, (22), 19–44.

Robic, M. C. (2006): Une Discipline se Construit. Enjeux, Acteurs, Positions. In Robic, M. C.; et al. (Eds.): *Couvrir le Monde. Un Grand* XXe *Siècle de Géographie Française*. Paris: ADPF, 15–52.

Robic, M. C.; Gosme, C.; Mendibil, D.; Orain, O.; Tissier, J. L. (Eds.) (2006): *Couvrir le Monde. Un Grand* XXe *Siècle de Géographie Française*. Paris: ADPF.

Sanders, L. (1984): *Interaction Spatiale et Modélisation Dynamique. Une Application au Système Intra-urbain*. State doctorial dissertation. Paris: University of Paris 1.

Soulié, C. (Ed.) (2012): *Un Mythe à Détruire? Origines et Destin du Centre Universitaire Expérimental de Vincennes*. Saint-Denis: Presses universitaires de Vincennes.

Soulié, C. (2013): Des Humanités à "l'économie de la Connaissance"? Les Transformations du Corps Enseignant en Lettres et Sciences Humaines en France (1949–2010). In Conesa, M.; et al. (Eds.): *Faut-il brûler les Humanités et les Sciences Humaines et Sociales?* Paris: Michel Houdiard eds.

Usselman, M. M. (2020): Strasbourg, May 1968. In Orain, O.; Robic, M.-C.: Mai-Juin 68, *L'Espace Géographique* et la Mémoire d'une Communauté, Special Issue. In *L'Espace Géographique*, 2020(1), 15.

8 How landscape became ecosystem

The nature of the quantitative revolution in German geography

Katharina Paulus and Boris Michel

Introduction

The history of the quantitative revolution in geography has received increased attention in recent years by a wide range of authors from a broad spectrum of geographical contexts. Beyond simplistic narratives of a radical break and a revolutionary paradigm shift, many of these authors highlight local specificities, continuities and blending of academic and the sociopolitical logics. Overall, these authors agree that the object and epistemology of geographical research changed fundamentally around the middle of the 20th century. The long-held view geography's task was to describe the total character of a terrestrial region lost its relevance and was replaced by the aim of explaining the spatial relations within the logics of mathematical theories and models. As the most prominent authors of the quantitative revolution as well as historians of the discipline are located on the human-geography side of the discipline, most of the research focuses on geography as social science and humanities.

This is problematic or at least one-sided. Taking the history of 20th-century German geography as an example, this paper argues that one often overlooked element of the quantitative revolution is how it relates to a change in the way, geographers thought about nature. While one of the dominant narratives of the history of the quantitative revolution takes the economic and urban geography of Walter Christaller (Christaller 1933) as a starting point for a new geography that is dominated by human geographers, we argue that it was also the introduction of a new way of thinking about nature that made this new geography work. Our argument is not that in addition to human geography we should also look at physical geography in order to have two separated stories, but that the concept of nature is relevant for the very existence of mid-20th-century geography as a whole. At least for German-speaking regional geography of the late 19th and early 20th century ("Länderkunde"),[1] the very existence of geography as an academic discipline relied upon its ability to understand itself as a bridge between nature and culture, between natural science and the humanities. We argue that the introduction of the concept of ecology and later the emergence of the ecosystem were ways to make a new geographical rationality plausible to geographers without tearing apart its holistic foundation. This new mode of

DOI: 10.4324/9781003122104-8

thinking about nature in terms of systems and cybernetics helped to modernize the old regional holism instead of calling for a radical break that – as many in German geography feared – could place geography as an academic discipline in jeopardy.

After a short introduction to the dominant narrative about the history of the quantitative revolution in German geography, this paper follows three concepts of nature – landscape, landscape ecology, and ecosystem – and their relation to the understanding of geography as a holistic discipline.

The quantitative revolution in the historiography of German geography

The dominant narrative about the quantitative and theoretical revolution in (West) German geography focuses on two events that took in the late 1960s when the quantitative revolution in some countries already started to be challenged by a new generation of radical geographers (Harvey 1972). The focus on these distinct moments not only tends to mythologize a long and complicated process by reducing it to singular events (Korf 2014) but also favours a story that concentrates on the social and cultural side of geography and leaves little room for physical geography and nature.

The Anglophone debates of the 1950s and 1960s were hardly received at the time and only had a limited impact on German geography. Much of the German geography after 1945 attempted to reconnect with an older regional geography that dominated before 1933 and only applied some minor functionalist modernizations (Bobek and Schmithüsen 1949), most notably in the small but emerging field of "social geography" (Bobek 1948). This was also a geography that – due to the role of spatial planning under National Socialism – shied away from a wholehearted embrace of applied geography and any risk of appearing "political" (Wardenga et al. 2011).

Instead of importing the Anglophone or international debate, it was the publication of one book in 1968 and the biennial meeting of German geographers in Kiel 1969, which mark the historical event of the quantitative and theoretical revolution in the discipline's collective memory (Wardenga 2019). In 1968, Dietrich Bartels published his "Zur wissenschaftstheoretischen Grundlegung einer Geographie des Menschen" (Bartels 1968), the first major attempt in German-speaking geography to formulate a theoretical framework for human geography that was based on scientific positivism and elements of critical rationalism. The book was both bold and difficult to read and in the preface the book series' editors distanced themselves from its content. To them, the book reads as a call for a separation of geography into two separated disciplines, one in the realm of the natural sciences and one in the social sciences, leaving little room for the integrative "Länderkunde". While this reading might be an exaggeration of two of the discipline's gatekeepers, Bartels formulates a mathematically systematically geography of spatial structures and he argues for geography as a theoretical and applied spatial science. The focus would then no longer be the description of landscapes as a tableau of nature and material culture, which makes the uniqueness of regions, but rather the search for universally valid spatial laws as theories. While Bartels is primarily interested in human geography and for him the link between natural science and social science was a unified methodology of science, he spends a considerable amount of time on current debates on ecology and ecosystems which he reads as a system-theoretical version of "Länderkunde" (Bartels 1968, 63–73).

The other historical event was the 1969 biennial Meeting of German Geographers in the city of Kiel. Today this event is best known for a paper delivered by a group of critical students that contains a broad critique of the state of geography in post-war Germany (Berliner Geographenkreis 1969). The students not only rejected the "pseudo-science" (Berliner Geographenkreis 1969, 13) of "Länderkunde" but also criticized the lack of social relevance of geography that follows a descriptive regionalist approach, both with regard to the changed situation in the labour market as well as with regard to critical approaches to understanding the crisis of late capitalism in the 1960.[2]

Since most historians of the discipline and most protagonists of the debates about theory and epistemology of the discipline positioned themselves closer to human geography than to physical geography, there is very little discussion about the role physical geography played.[3] Thus, it is not surprising that the narratives about this paradigm shift focus on those who – like Walter Christaller and Dietrich Bartels – modernized social, urban and economic geography. Overall, this is the story of a quantitative and theoretical revolution in geography that took place in human geography. While Bartels and others saw the link between human geography and physical geography in a unified methodology of science, large parts of the traditional geography rejected this new geography for – amongst other things – the way it separated human geography from physical geography. In this narrative, the field of physical geography is largely absent and appears undisturbed by these debates, which in turn seems to have been of little concern to historians of geography.

From landscape to landscape ecology

The term ecology entered geography more than half a century after German biologist Ernst Haeckel had first introduced it. He used it in 1866, to describe the relationship an organism has with the surrounding environment (Haeckel 1866, 286). At that time, the concept of ecology was strongly limited to the connections between individual organisms and their environment (Möbius 1877; Brandt 1899; Thienemann 1918). Ecology developed into a more comprehensive concept in the period around the First World War and was now used to describe the coexistence of all living organisms within one location. After that, the biological concept was transferred into other spheres and disciplines. This reached from being applied to other natural sciences to being used to describe the human and social world in the social sciences where it laid the foundation of the idea of social ecology that entered into Anglophone geography in the 1920s (Barrows 1923).

While there is not enough space in this paper to elaborate more extensively on the concept of nature in the 19th- and early 20th-century German geography, it is worthwhile to highlight a few points. Most historiographic discussions about the concept of nature in the early 20th century centre around the question of environmental determinism, but it is also important to emphasize the way nature is primarily perceived in its visual and tangible form. Natural landscape, Carl Troll wrote as late as 1950, "is understood as a part of the earth's surface that through its appearance and the combination of its phenomena as well as its inner and outer locations forms a spatial unit with a distinct character" (Troll 1950, 165). Here nature is a scenic and total landscape and perceived as an assemblage of collectible things in space. Geographers encounter this nature in the field and not in the laboratory. It was this concept of "landscape" that was the centre of the 1960s critique. But even before these debates arose, the concept of landscape ecology began to change the understanding of nature in German geography.

Landscape ecology

In 1939, the German geographer Carl Troll[4] published a much-noticed and frequently quoted article with the lengthy title "Luftbildplan und ökologische Bodenforschung. Ihr zweckmäßiger Einsatz für die wissenschaftliche Erforschung und praktische Erschließung wenig bekannter Länder" (Aerial photo map and ecological soil research. Their appropriate use for scientific research and practical development of little-known countries). Therein he coined the term "Landschaftsökologie" (landscape ecology) which knitted together the new term ecology that came from a different discipline – and was used by Troll without any reference to human or social ecology – with the cornerstone of hegemonic geography, the concept of "Landschaft" (landscape).

Even though his 1939 paper was later mostly referred to in the context of early ecological research (Finke 1971; Müller 1974; Turner 1989; Steinhardt 2012), its main argument was to highlight the advantages of the emerging field of aerial photography over conventional methods of geographical research (Troll 1939, 241). Troll's interest in aerial photography dates back at least to his research in the Andes and to his participation in the third German Nanga Parbat excursion in 1937.[5] In both of these research projects, Troll was able to use airplanes to generate photographic information. He thereby supports Peter Anker who argues with regard to British work on ecology that "the most important new technology in the enlargement of ecological method was the airplane" (Anker 2001, 3). For Troll, aerial photography was less an easy and cheap way to create maps – preferably in colonial contexts or in what the article's title refers to as "little known countries" – than a new way of looking at things. Aerial photography, Troll writes, "preserves the image and visual content of lands and landscapes" (Troll 1939, 242) and – against the tendency of modern sciences to specialize and divide its objects – helps geography to keep its focus on the synthesis of nature and culture in the cultural landscape. Thereby this modern technological innovation promises to keep geography intact as a discipline that began to cope with a growing diversification between its subfields and neighbouring disciplines. The notion of ecology as a holistic perspective on nature and a perspective that engages with nature in the field, not in a laboratory, provides a fitting new term for an old geographical topos. But more than being a new name for an old concept, ecology provided for – what we today would call – interdisciplinary research. This becomes especially evident in the context of the German war effort.

In order to obtain the necessary financial means for his research after the war began, Troll tied his aerial photography to an existing project of a research group headed by geographer Otto Schulz – Kampfhenkel, which was considered vital for the war. This research squadron ("Forschungsstaffel Schulz-Kampfhenkel") was founded in 1940 and turned into a special unit of the German Foreign Intelligence Service in 1943. It had the task to use aerial photographs to assess the terrain and produce maps for combat troops by combining the aerial photographs with classical geographical knowledge. In 1944, Troll was commissioned by the Reich Research Council to improve the geographical analysis of aerial photographs and to translate Russian literature on remote sensing for the research squadron.[6] Troll himself was not an employee of the research squadron, but his research assignment with the unit prevented him from being drafted and provided him with access to the research squadrons archive of aerial photographies (Fahlbusch et al. 1989; Häusler 2007, 7, 182–183; Flachowsky, 272–274). The research team consisted not only of geographers but also, depending on the nature of the task, of cartographers, geologists, soil scientists, vegetation scientists, meteorologists, astronomers, hydrologists and road engineers, amongst them some of the future key protagonist of ecosystem theory such as Heinz Ellenberg. Within the interdisciplinary research squadron Schulz Kampfhenkel, the use of aerial photography in wartime proved to be essential, but cooperation also had an effect on the further development of landscape ecology after the war (Böhm 1995, 130, 137; Häusler 2007, 182–183).[7]

Research following Troll

The small number of research projects and publications in geography that drew on the concept of ecology after the war and until the late 1950s were often strongly influenced by Troll's idea of landscape ecology. Although they also put vegetation and animals that characterize an area in relation to each other, they did not draw quantifiable or systematic conclusions from them, as the emerging field of ecosystem research increasingly did at that time (Kwa 2018). Geographical research in Germany continued to focus on debates on classifying landscapes and in general used the term "Landschaftsökologie" primarily to talk about landscape (Bobek and Schmithüsen 1949). While this term was used in a slightly more analytic way and linked to a language that was more akin to the language of modern science than the often deeply romantic or descriptive language of landscape geography, the difference lay more in terminology than epistemology. By focusing on the determination of natural spaces, the discipline retained its focus without completely abandoning holistic research. Some of the ideas around the concept of ecology were translated into the logics of German regional studies without forcing a complete break with the old concept.

Unsurprisingly, German geography returned to a concept of landscape that focused on physiognomy, on visual perception of nature and on typologies. One of the most ambitious projects of the academic geography from the early 1950s onwards was a handbook that would give a comprehensive account of the natural structuring of Germany ("Handbuch der naturräumlichen Gliederung Deutschlands (Meynen and Schmithüsen 1953–1962)). Since almost all geographical institutes were involved in this research project, it was inevitable that there would mainly be disputes over uniform designations and scales and not about the usefulness of the project and the discipline in general.

Landschaftsökologie revisited

After the first article, Troll himself published little on the subject of landscape ecology. At the same time, the field of ecological research underwent fundamental development, and researchers increasingly focused on ecosystem research and linked their considerations closely to the development of the first computers.

Troll reviewed other publications on this subject and argued against plant sociology, which he regarded as an artificial classification, whereas ecological investigations would lead to more natural harmonic results (Troll 1956; Kwa 2018, 184). In general, ecology did not become a centrepiece of his work. This is also reflected in a ceremonial address to Troll, given in 1959 by a colleague, where the concept of landscape ecology is not mentioned at all (Lautensach 1959).

It is only in the mid-1960s in a contribution to a conference of biological vegetation researchers that Troll returns to the concept (Troll 1968b). He stays true to his earlier paper but reframes it much more within the now growing debate, as a central field of geographical research. While the new interpretation still centres on landscape, the main task is now to subject these to a functional analysis that draws on the new language of the ecosystem and identifies the interdependencies between plants and their environment. Troll aims at showing that landscapes are not merely subjective or aesthetic, but an expression of their inner essence and thus objective or physiognomic (Kirchhoff 2011, 53). To Troll, it is a specific, particularly comprehensive approach to the spatially limited effects of plants and animals and their effects on human activities.

In a way, Troll's work from this period complements the 1939 article: Now, landscape ecology is the central object of investigation and aerial photography is merely the method (Troll 1939, 1968b, 1968a, 1972). The change in perspective and terminology can be explained by Troll's greater involvement in the international debates on ecosystems from the late 1960s onwards. The most far-reaching change is certainly the translation of landscape ecology with the term geoecology. However, landscape is still present conceptually, even though it has disappeared in name. Thus in his obituary written in 1976, "Landschaftsökologie" or geoecology is given a central position and is even described as "a methodical basic framework, a model of thought [that] repeatedly [returns] in the other fields of work of C. Troll" (Lauer 1976). What becomes evident in these changed perspectives within Troll's biography is an indication of the changes affecting the concept of ecosystem throughout the discipline of geography.

In the 1960s, most German geographers still saw themselves as classic landscape geographers and the number of papers explicitly dealing with different aspects of ecology and the ecosystem is relatively low. The increase of papers published in German geographical journals like "Erdkunde", "Geographische Zeitschrift", "Die Erde", and the Swiss paper "Geographica Helvetica" between 1950 and 1979 (Figure 8.1) on this subject is an indication for the wide ranging changes the discipline underwent in the 1970s (458 papers mention "Ökologie" and 270 the English term "ecology", there are also 132 mentioning "Landschaftsökologie", 55 referring to "Ökosystem" and 34 to "ecosystem"). The term Ökologie was originally mainly used in vegetation geographic papers and is only later picked up in papers in human geography and in theoretical considerations, but in these cases it is more likely that ecology is placed in the broader context of ecosystem and system theory.

In Germany, the geobotanist and biologist Heinz Ellenberg[8] was one of the central figures in early ecosystem research. Ellenberg's work was much quoted and discussed by geographers: "Ellenberg's book is likely to be the standard

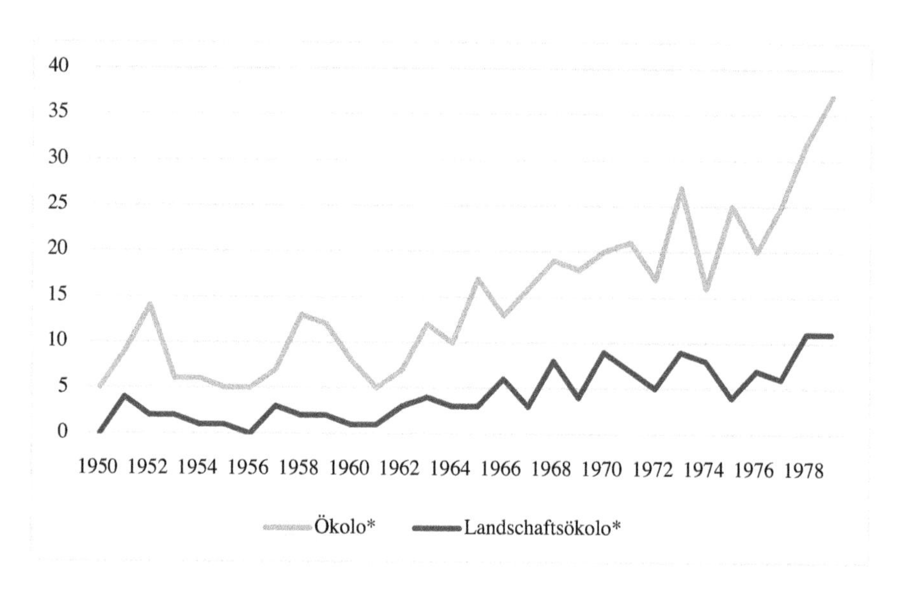

Figure 8.1 Number of articles using the terms "Ökologie" (including all variations) and "Landschaftsökologie" (including all variations) in four of the main German geographical journals between 1950 and 1979 (own research).

work in Central Europe's general vegetation geography for a long time to come" (Troll 1966). This close cooperation between geography and biology allowed Ellenberg's interpretation of the ecosystem to shape both disciplines (Fliedner 1979). Between 1966 and 1973, he headed the UNESCO funded Solling project aimed at gaining insights into complex ecosystem modelling (Ellenberg et al. 1986). The project was part of the UNESCOs international biological program (IBP), a large-scale research initiative with the purpose of accumulating data-sets of various complex ecosystems. The Solling project was one of the most comprehensive European contributors but compared to the American grassland biome project at the University of Colorado set out to develop a total ecosystem mode, it was on a small scale. The Grassland project can be considered the first big science project in biology and intended to produce a complete model of an ecosystem, but the researchers encountered massive problems in combining and making use of the data they produced (Innis et al. 1978; Kwa 1993; Golley 1993). The German Solling project also encountered problems with the processing of collected data but the project was less focused on comprehensive models and quantitative data and the authors explicitly criticized the IBP-projects that aimed at comprehensive and predictive models of the ecosystem (Ellenberg et al. 1986, 28, 31). Ellenberg does not see the ecosystem simply as the sum of its elements but as a unity or wholeness and sees the actual task of ecosystem research in understanding the system as a whole. Regardless of this, he does not use the metaphysical vocabulary that is often used by holistic landscape geography but applies the language of system theoretical concepts (Jax 1998, 137). Therefore, Ellenberg's interpretation of the ecosystem allows geography to go on without a complete dissolution of the classical research ideas. In seeing the ecosystem as a contemporary version of the landscape, it could be integrated into the existing epistemologies of geography and thereby paved the way for the less holistic systems-theoretical ecology research of Anglo-American geography.

Ecosystem or the concept of nature in the "limits of growth" era

In the course of the 1960s, both general systems theory and cybernetics gained nationally and internationally in importance in science as well as in society and offered methodical approaches that promised to help understand the limitations humanity was confronted with. From today's perspective, it seems as if the view of nature had changed fundamentally in the early 1970s maybe best exemplified by the 1972 report "The limits to growth" (Meadows et al. 1972) of the Club of Rome. The study was based on cybernetic modelling and Jay Forrester's concept of "world system". As a consequence of this publication, nature was now widely accepted as limited and brought out of natural equilibrium and possibly heading for collapse. But nature was also understood as something that could be described and predicted by computer models. Only through the translation of computers was it possible to understand the consequences of human actions on nature (Schmelzer 2017).

This consequently also changed the way geography in Germany dealt with the concept "ecosystem". The rapid technological progress increased the quantity and quality of the data available for research and ecosystem research required, as the Solling and Grasslands projects proved, large amounts of structured data to determine the connections and resilience of system components. Improved data quality established, for example, a higher level of detail, making it possible to use the data for complex statistical analysis. In geography, the availability of geoinformation data generated by satellites since the late 1960s, together with machine-readable data sets on population and environment, allowed for fundamentally changed research practices.

Physical geography integrated the system theoretic view of the ecosystem more easily due to higher specialization within the different branches and developments in neighbouring disciplines with whom some researchers worked closely together. In the 1960s, ecosystem research in physical geography was still very clearly oriented towards vegetation-geographical questions and was only later on also used in other contexts (Klink 1964). This resulted in a number of publications in the 1970s where nature was seen and analysed through the view of systems analysis (Walter 1971).

The path of systems theory into human geography was somewhat more intricate. Those remembered today as reformers of the discipline mainly advocated a positivist and critically rationalist approach and treated questions of system theory rather subordinately. Dietrich Bartels, read Chorley, Stoddard, and other Anglo-American Geographers who were working towards a systems theory perspective in geography. With his anthology "Economic and Social Geography", he was the first to publish a translation of Stoddard's "Geography and the ecological approach. The geosystem as a Geographic principle and method" (Stoddart 1970) into German. But in the editorial he deliberately distances himself from a structure completely relying on systems theory, since he considers both the theory and the human understanding to be overburdened by it (Bartels 1970). Consequently Bartels opts in his "Geographie des Menschen" for a human geography that is clearly separate from physical geography but promotes a critical rationalist approach and quantitative methods without a system theoretical perspective in mind (Bartels 1968).

A number of geographers interested in systems theory can be found in the remaining protagonists of the classical regional geography. Here systems theory is described as a way out for the discipline that maintains the holistic paradigm and provides a modern language for "Länderkunde". The best example for this approach might be the book "Theoretische Geographie" (theoretical geography) by Eugen Wirth in which he elaborately connects systems theoretical approaches with regional geography (Wirth 1979). A work that was sharply criticized for varying reasons, one being its embracing of new theoretical approaches into the logics of regional geography (Bartels 1980; Dürr 1979; Hard 1986).

But also, from the ranks of younger geographers who are not primarily interested in the rehabilitation of the regional paradigm, attempts are being made to make use of the integrating potentials of the ecosystem approach within geographical research. Particularly noteworthy in this context is the dissertation by Peter Weichhart (Weichhart 1975), which was perceived by many as a response to Bartels "Geographie des Menschen" (Bartels 1968). Weichhart opts for a geography partly united under the common idea of a system-theoretical ecology which devotes itself to complex interrelationships of effects.

> The ecological idea is above all a monistic concept. By considering the interrelations and interdependencies between society and the physical environment, the concept of ecology closes the gap between nature and intellect and thus also eliminates the alleged dichotomy between physiogeography and anthropogeography. Society and the environment are understood together by a system of concepts, and with the help of a methodological concept they are seen as a unit, namely as an "ecosystem".
>
> (Weichhart 1975, 89 translated by K. Paulus)[9]

Weichhart was criticized for his call for a unified geography and even put on the spot for enabling the return of the "Länderkunde" (Hard 1975) through the terminologies of the ecosystem. So even though the conflict was not yet settled and he later clarified that he does not expect all subfields to gather under the ecosystem approach but wants to create a bridge that allows them to work together on questions related to human–environment interactions (Weichhart 1976), the debates about the nature of geography became less central. After that, ecosystem approaches were still part of both sub-disciplines but they had lost the appeal of a possible return to an integrated geography with a unified epistemology.

Conclusion

One of the most important developments in the natural sciences of the post-war period was the fundamental change in the concept of nature through the emergence of the ecosystem concept. But although the ecosystem idea as an approach to understanding nature became hegemonic at that time, the massive changes that took place at the same time in geographical research are rarely associated with it, even though the discipline's self-understanding was to describe and analyse nature.

In traditional landscape geography, which was still very widespread in German geography in the 1960s, human action is seen as shaped or even determined by the surrounding landscape. At the same time, the natural sciences have increasingly viewed nature as a self-contained system in which human action is only of interest, if it interferes with the balance of nature.

However, while most narratives of the quantitative turn in geography now assume a clear break between these phases, we argue in our contribution that the idea of the ecosystem coincided with many assumptions of the classical landscape approach and that these were initially translated into the language of system theory and were therefore known when used in the applied and specialized sub-disciplines. Accordingly, the assumed clean break between completely contrary approaches is neither so clean nor so contrary. The concept "ecosystem" did not question the foundations of geography and the revolutionary aim to create a model of an ecosystem where every aspect of nature would be synthesized was something, they were already familiar with. This offers classical landscape geographers a bridge that could be built by the two concepts inherent wholeness. Most geographers were already doing "Ganzheitsforschung" (wholeness research) and some were (with different degrees of success) able to combine this with the ideas of systems theory.

But the process of two sub-disciplines drifting apart, irreversibly establishes two separate disciplines. The publication of Weichhart's dissertation in 1975 is perceived by many as a last attempt not to let these parts drift apart completely (Giese 1977). An intervention that neither calls for a return to a holistic geography nor does it consider the complete separation of the sub-areas as demanded by Bartels as the end point. But the empirical objects: Peoples, societies, climate, behaviour, vegetation, cities, cultures, and glaciers are no longer part of each other's research scope.

The general focus on the significance of Christaller's new approach of seeing landscape with the means of centrality research makes one forget that approaches such as a system-theoretically understood ecosystem managed to bridge the gap between because they were connectable to both theoretical approaches.

Notes

1 In our paper, we use "Länderkunde" when we refer to the dominant paradigm of German geography between the late 19th and mid-20th century. The terms "regional geography" or "Landschaftsgeographie" describe the same paradigm that is closely linked with the work of Carls Sauer (Sauer 1925) in Anglophone geography.

2 In recent years, new research made this story more porous. Researchers started to tell more nuanced histories that focus on local variations, and prehistories as well as the slow and contradictory process of establishing quantitative and theoretical geography in Germany (Steinkrüger 2015; Kemper 2018; Michel and Paulus 2018). The research of Belina et al. 2017 and others has shown that the role of critical geography in the memory of the discipline remains marginal to this day. The potential for change is mainly attributed to the criticism of the scientific foundation of the discipline and of the antiquated methods. The more fundamental points of criticism of political and economic systems and National Socialist continuities have been neglected at the time and are still underrepresented today. The history of applied research (Wardenga et al. 2011) shows that although this branch of research was also excluded from mainstream geography in its time of emergence it later gained a leading role in the discipline.

3 A notable exception would be Gerhard Hard, a vegetation-geographer and one of the most vocal critics of the concept of "Länderkunde" (Hard 1965, 1969, 1975).

4 Troll, who had done extensive work on plant geography in the Andes in the 1920s, was appointed Professor of Geography at the University of Bonn in 1938. He took over the chair of Leo Waibel who had previously been forced out of academia by the Nazi regime and later emigrated to the US (Schmithüsen 1952; Baas 2015). While not a party member and keeping some distance to the regime, Troll made a successful career under National Socialism. After 1945, he became one of the leading German geographers and president

of the IGU in the 1960s. In 1947, he published the defining and exonerating paper on German geography during national socialism (Troll 1947; Michel 2016).

5 The Nanga Parbat excursion was publicly announced as the heroic conquest of the "Schicksalsberg der Deutschen" (The Fate Mountain of the Germans). The reporting was exclusively done by the Nazi party's newspaper "Völkischer Beobachter", portraying it as the prototypical National Socialist heroic male undertaking. It ended with the death of 16 people in an avalanche and was the biggest mountaineering accident in the Himalaya at the time (Mierau 2006, 194f.). Troll later claimed that he never felt comfortable with this heroism and the pressure to join the NSDAP after he returned to gain an important scientific position in Berlin (Kwa 2018; 194; Böhm and Aymans 1991, 245).

6 For the translation of the Russian literature, Troll hired the Innsbruck botanist Helmut Gams, who had been discharged from military service because he could not present proof of Aryan origin. With the special powers of his position, von Schulz-Kampfhenkel was able to employ Gams as a "civilian" in the research squadron (Flachowsky 2011, 272; Häusler 2007, 83).

7 Due to its interdisciplinary structure, the significance of the Schulz-Kampfhenkel research squadron can in many ways be paralleled with American war research, which Barnes describes as one of the central organizational moments for quantitative geography in the USA (Barnes 2015, 2013; Barnes and Crampton 2011).

8 Ellenberg wrote a pioneering synecological doctoral thesis in 1939 (Ellenberg 1939) followed by interdisciplinary cooperation and research on early questions of ecology. Between 1943 and 1945, Ellenberg was employed at the research squadron Schulz-Kampfhenkel (Häusler 2007, 170) where he focused his research on botanically applied questions (Dierschke 1997, 6). He oversaw several projects concerning the production of maps for terrain assessment and tank maps for the eastern front (Häusler 2007, 72).

9 Original in Weichhart 1975, 89:

> Der Ökologiegedanke ist vor allem aber ein monistisches Konzept. Indem die Interrelationen und Interdependenzen zwischen Gesellschaft und physischer Umwelt betrachtet werden, schließt das Ökologiekonzept die Kluft zwischen Natur und Geist und beseitigt damit auch die angebliche Dichotomie zwischen Physiogeographie und Anthropogeographie. Gesellschaft und Umwelt werden gemeinsam durch ein Begriffssystem erfaßt [sic], mit Hilfe einer methodologischen Konzeption als Einheit gesehen, nämlich als "Ökosystem".

References

Anker, P. (2001): *Imperial Ecology: Environmental Order in the British Empire, 1895–1945.* Cambridge, MA: Harvard University Press.

Baas, K. (2015): *Erdkunde als politische Angelegenheit: Geographische Forschung und Lehre an der Universität Münster Zwischen Wissenschaft und Politik (1909–1958).* Münster: Aschendorff Verlag.

Barnes, T. (2013): Folder 5, box 92. In *RSCG*, 14(7), 784–791.

Barnes, T. (2015): War by Numbers: Another Quantitative Revolution. In *Geopolitics*, 20(4), 736–740. DOI: 10.1080/14650045.2015.1095588.

Barnes, T.; Crampton, J. W. (2011): Mapping Intelligence, American Geographers and the Office of Strategic Services and GHQ/SCAP (Tokyo). In Kirsch, S.; Flint, C. (Eds.): *Reconstructing Conflict: Integrating War and Post-War Geographies.* Farnham, Surrey, Burlington, VT: Ashgate.

Barrows, H. H. (1923): Geography as Human Ecology. In *Annals of the Association of American Geographers*, 13(1), 1–14. DOI: 10.1080/00045602309356882.

Bartels, D. (1968): *Zur wissenschaftstheoretischen Grundlegung einer Geographie des Menschen.* Wiesbaden: Steiner.

Bartels, D. (Ed.) (1970): *Wirtschafts- und Sozialgeographie.* Köln [u.a.]: Kiepenheuer & Witsch.

Bartels, D. (1980): Die konservative Umarmung der "Revolution". Zu Eugen Wirths Versuch in "Theoretischer Geographie". In *Geographische Zeitschrift*, 68(2), 121–131.

Belina, B.; Strüver, A.; Naumann, M.; Best, U. (2017): Better Late Than Never? Critical Geography in German Speaking Countries. In Berg, L. D.; Best, U. (Eds.): *Placing Critical Geography: International Histories of Critical Geographies.* London: Routledge, 1st edition.

Berliner Geographenkreis (Ed.) (1969): Sonderheft zum 37. deutschen Geographentag. In *GEOgrafiker*, (3).

Bobek, H. (1948): Stellung und Bedeutung der Sozialgeographie. In *Erdkunde*, 2, 118–125.

Bobek, H.; Schmithüsen, J. (1949): Die Landschaft im logischen System der Geographie. In *Erdkunde*, 3(2/3), 112–120.

Böhm, H. (1995): Luftbildforschung. Wissenschaftliche Überwinterung – Angewandte (kriegswichtige) Forschung – Rettung eines Paradigmas. In Wardenga, U.; Hönsch, U. (Eds.): *Kontinuitäten und Diskontinuitäten der deutschen Geographie in Umbruchphasen.* Münster: Institut für Geographie der Westfälischen Wilhelms-Universität, 129–140.

Böhm, H.; Aymans, G. (Eds.) (1991): *Beiträge zur Geschichte der Geographie an der Universität Bonn.* Herausgegeben anläßlich der Übergabe des neuen Institutsgebäudes in Bonn-Poppelsdorf. Bonn: Dümmler in Komm (Dümmlerbuch, 7421).

Brandt, K. (1899): *Über den Stoffwechsel im Meere.* Kiel: Rede beim Antritt des Rektorates der königlichen Christian-Allbrechts-Universität zu Kiel.

Christaller, W. (1933): *Die Zentralen Orte in Süddeutschland.* Jena: Gustav Fischer Verlag.

Dierschke, H. (1997): Heinz Ellenberg. (1913–1997). In *Tüxenia*, 17, 5–10.

Dürr, H. (1979): *Für eine offene Geographie, Gegen eine Geographie im Elfenbeinturm.* Karleruhe: Geographisches Institut Universität Karlsruhe (Karlsruher Manuskripte zur mathematischen und theoretischen Wirtschafts- und Sozialgeographie, 36).

Ellenberg, H. (1939): *Über Zusammensetzung, Standort und Stoffproduktion bodenfeuchter Eichen- und Buchen-Mischwaldgesellschaften Nordwestdeutschlands.* Hannover: Engelhard (Mitteilungen der floristisch-soziologischen Arbeitsgemeinschaft in Niedersachsen).

Ellenberg, H.; Mayer, R.; Schauermann, J. (Eds.) (1986): *Ökosystemforschung. Ergebnisse des Sollingprojekts; 1966–1986.* Stuttgart: Ulmer.

Fahlbusch, M.; Rössler, M.; Siegrist, D. (Eds.) (1989): *Geographie und Nationalsozialismus. Drei Fallstudien zur Institution Geographie im Deutschen Reich und der Schweiz.* Kassel: Gesamthochschule (Urbs et regio, 51).

Finke, L. (1971): Landschaftsökologie als angewandte Geographie. In *Berichte zur deutschen Landeskunde*, 45, 167–182.

Flachowsky, S. (2011): Die Forschungsgruppe Schulz-Kampfhenkel steht jetzt für Ostaufgaben zur Verfügung. In Stoecker, H.; Flachowsky, S. (Eds.): *Vom Amazonas an die Ostfront. Der Expeditionsreisende und Geograph Otto Schulz-Kampfhenkel (1910–1989).* Köln/Wien: Böhlau Verlag, 240–302.

Fliedner, D. (1979): Geosystemforschung und menschliches Verhalten. In *Geographische Zeitschrift*, 67, 29–42.

Giese, E. (1977): Book Review. Weichhart, Peter: Geographie im Umbruch. Ein Methodologischer Beitrag zur Neukonzeption der komplexen Geographie. In *Geographische Zeitschrift*, 65(2), 146–147.

Golley, F. B. (1993): *A History of the Ecosystem Concept in Ecology: More Than the Sum of the Parts.* New Haven, CT: Yale University Press.

Haeckel, E. (1866): *Generelle Morphologie der Organismen. Bd. 2: Allgemeine Entwickelungsgeschichte der Organismen: Kritische Grundzüge der mechanischen Wissenschaft von den Entstehenden Formen der Organismen.* Berlin: Reimer.

Hard, G. (1965): Arkadien in Deutschland. Bemerkungen zu einem landschaftlichen Reiz. In *Die Erde*, 96, 21–41.

Hard, G. (1969): Die Diffusion der "Idee der Landschaft": Präliminarien zu einer Geschichte der Landschaftsgeographie (The Diffusion of the Idea of the Landscape). In *Erdkunde*, 23(4), 249–264.

Hard, G. (1975): Von der Landschafts- zur Ökogeographie. Zu den methodologischen Überlegungen von Peter Weichhart. In *Mitteilungen der Österreichischen Geographischen Gesellschaft*, 117, 274–286.

Hard, G. (1986): Der Raum – einmal systemtheoretisch gesehen. In *Geographica Helvetica*, 41(2), 77–83.

Harvey, D. (1972): Revolutionary and Counter Revolutionary Theory in Geography and the Problem of Ghetto Formation. In *Antipode*, 4(2), 1–13.

Häusler, H. (2007): Forschungsstaffel z.b.V. Eine Sondereinheit zur Militärgeographischen Beurteilung des Geländes im 2. Weltkrieg. Wien (MILGEO – Schriftenreihe des Militärischen Geowesens, 21).

Innis, G. S.; et al. (Eds.) (1978): *Grassland Simulation Model*. New York: Springer.

Jax, K. (1998): Holocoen and Ecosystem: On the Origin and Consequences of Two Concepts. In *Journal of the History of Biology*, 31, 113–142.

Kemper, J. (2018): Fortschritt und Verdrängung. Ökologischer Fehlschluss und quantitative Revolution in der Geographie. In *Geographica Helvetica*, 73(1), 49–61. DOI: 10.5194/gh-73-49-2018.

Kirchhoff, T. (2011): Landschaftsökologie gleich Ökologie der Landschaft? Eine wissenschaftstheoretisch-kulturwissenschaftliche Analyse landschaftsökologischer Forschungsprogramme. In *Laufener Spezialbeiträge*, 53–60.

Klink, H. (1964): Landschaftsökologische Studien im südniedersächsischen Bergland (Regional Ecological Studies of the Hill Country of Southern Lower Saxony). In *Erdkunde*, 18(4), 267–284.

Korf, B. (2014): Kiel 1969 – Ein Mythos? In *Geographica Helvetica*, 69(4), 291–292. DOI: 10.5194/gh-69-291-2014.

Kwa, C. (1993): Modeling the Grasslands. In *Historical Studies in the Physical and Biological Sciences*, 24(1), 125–155.

Kwa, C. (2018): The Visual Grasp of the Fragmented Landscape: Plant Geographers vs. Plant Sociologists. In *Historical Studies in the Natural Sciences*, 48(2), 180–222. DOI: 10.1525/hsns.2018.48.2.180.

Lauer, W. (1976): Carl Troll – Naturforscher und Geograph. In *Erdkunde*, 30(1). DOI: 10.3112/erdkunde.1976.01.01.

Lautensach, H. (1959): Carl Troll – ein Forscherleben. In *Erdkunde*, 13(4). DOI: 10.3112/erdkunde.1959.04.01.

Meadows, D. H.; Meadows, D.; Randers, J.; Behrens, W. (1972): *The Limits to Growth: A Report for the Club of Rome's Project on the Predicament of Mankind*. New York: Universe Books.

Meynen, E.; Schmithüsen, J. (Eds.) (1953–1962): *Handbuch der naturräumlichen Gliederung Deutschlands. Band 1 1953–1962. Unter Mitwirkung des Zentralausschusses für deutsche Landeskunde*. Bad Godesberg: Bundesanstalt für Landeskunde und Raumforschung.

Michel, B. (2016): "With Almost Clean or at Most Slightly Dirty Hands": On the Selfdenazification of German Geography after 1945 and Its Rebranding as a Science of Peace. In *Political Geography*, 55, 135–143.

Michel, B.; Paulus, K. (2018): Jenseits von Kiel: Zu einer Wissenschaftsgeschichte der quantitativ-theoretischen Wende in der deutschsprachigen Geographie. In *Geographica Helvetica*, 73, 301–307.

Mierau, P. (2006): *Nationalsozialistische Expeditionspolitik. Deutsche Asien-Expedition 1933–1945*. München: Utz (Münchner Beiträge zur Geschichtswissenschaft, 1).

Möbius, K. (1877): *Die Auster und die Austernwirthschaft*. Berlin: Wiegandt Hempel & Parey.

Müller, P. (1974): Was ist "Ökologie"? In *Geoforum*, 5(2), 78–81. DOI: 10.1016/0016-7185(74)90011-6.

Sauer, C. (1925): The Morphology of Landscape. In *University of California Publications in Geography*, 2(2), 19–54.

Schmelzer, M. (2017): "Born in the Corridors of the OECD": The Forgotten Origins of the Club of Rome, Transnational Networks, and the 1970s in Global History. In *Journal of Global History*, 12(1), 26–48. DOI: 10.1017/S1740022816000322.

Schmithüsen, J. (1952): Leo Waibel. 22.1. 1888 bis 4. 9. 1951. In *Die Erde*, 83, 99.

Steinhardt, B.; et al. (2012): Landschaft als Gegenstand wissenschaftlicher Erkenntnis. In Steinhardt, U.; Barsch, H.; Blumenstein, O. (Eds.): *Lehrbuch der Landschaftsökologie*. 2. überarbeitete und ergänzte Auflage. Heidelberg: Spektrum Akademischer Verlag, 23–69.

Steinkrüger, J.-E. (2015): Jenseits der Gründungsmythen – Kiel und die historische Geographie. In *Geographica Helvetica*, 70(3), 251–254.

Stoddart, D. R. (1970): Die Geographie und der ökologische Ansatz. In Bartels, D. (Ed.): *Wirtschafts- und Sozialgeographie*. Köln [u.a.]: Kiepenheuer & Witsch (Neue wissenschaftliche Bibliothek, 35), 115–124.

Thienemann, A. (1918): Lebensgemeinschaft und Lebensraum. In *Naturwissenschaftliche Wochenschrift*, 17(33), 281–290.

Troll, C. (1939): Luftbildplan und ökologische Bodenforschung. Ihr zweckmäßiger Einsatz für die wissenschaftliche Erforschung und praktische Erschließung wenig bekannter Länder. In *Zeitschrift der Gesellschaft für Erdkunde zu Berlin*, 1939(7, 8), 241–298.

Troll, C. (1947): Die geographische Wissenschaft in Deutschland in den Jahren 1933 bis 1945. In *Erdkunde*, 1(1/3), 3–48.

Troll, C. (1950): Die geographische Landschaft und ihre Erforschung. In Bauer, K. H.; Curtius, L.; Einem, H.; Ernst, F.; Friedrich, H.; Fucks, W.; et al. (Eds.): *Studium Generale. Zeitschrift für die Einheit der Wissenschaften im Zusammenhang ihrer Begriffsbildungen und Forschungsmethoden*. Berlin and Heidelberg, Springer, 163–181.

Troll, C. (1956): Das Pflanzenkleid der Tropen in seiner Abhängigkeit von Klima, Boden und Mensch. In *Geographische Berichte*, 1, 21–35.

Troll, C. (1966): Das Pflanzenkleid Mitteleuropas: Zu Heinz Ellenbergs "Vegetation Mitteleuropas mit den Alpen". In *Erdkunde*, 20(4), 303–305.

Troll, C. (Ed.) (1968a): Geo-Ecology of the Mountainous Regions of the Tropical Americas: With 10 Tables. Proceedings of the Unesco Mexico Symposium, August, 1–3, 1966, Organized under the Sponsorship of the Unesco Natural Resources Research Division in Connection with the Latin America Regional Conference, LARC, of the International Geographical Union. Bonn: Dümmler (Colloquium Geographicum).

Troll, C. (1968b): Landschaftsökologie. In Tüxen, R. (Ed.): *Pflanzensoziologie und Landschaftsökologie*. Dordrecht: Springer (Berichte über die Internationalen Symposia der Internationalen Vereinigung für Vegetationskunde, 7), 1–21.

Troll, C. (Ed.) (1972): Geoecology of the High-Mountain Regions of Eurasia. Proceedings of the Symposium of the International Geographical Union, Commission on High-Altitude Geoecology, November 20–22, 1969 at Mainz in Connection with the Akademie der Wissenschaften und der Literatur in Mainz. Kommission für erdwissenschaftliche Forschung. Wiesbaden: Steiner (Erdwissenschaftliche Forschung).

Turner, M. G. (1989): Landscape Ecology: The Effect of Pattern on Process. In *Annual Review of Ecology and Systematics*, 20(1), 171–197. DOI: 10.1146/annurev.es.20.110189.001131.

Walter, H. (1971): Biosphäre, Produktion der Pflanzendecke und Stoffkreislauf in ökologisch-geographischer Sicht. Mit 1 Abb./Biosphere, Production of the Plant Cover and Metabolism in Ecological-geographical Perspective. In *Geographische Zeitschrift*, 59(2), 116.

Wardenga, U. (2019): Vergangene Zukünfte – Oder: Die Verhandlung neuer Möglichkeitsräume in der Geographie. In *Geographische Zeitschrift*, 1. DOI: 10.25162/gz-2019-0009.

Wardenga, U.; Henniges, N.; Brogiato, H. P.; Schelhaas, B. (2011): *Der Verband deutscher Berufsgeographen 1950–1979. Eine sozialgeschichtliche Studie zur Frühphase des DVAG.* Leipzig: Leibniz-Institut. für Länderkunde (Forum IfL, 16).

Weichhart, P. (1975): *Geographie im Umbruch: ein methodologischer Beitrag zur Neukonzeption der komplexen Geographie.* Wien: Deuticke.

Weichhart, P. (1976): Anmerkungen zum Dogma der uneinigen Geographie. Gerhard Hards Kritik an der Ökogeographie. In *Mitteilungen der Österreichischen Geographischen Gesellschaft*, 118(1), 195–208.

Wirth, E. (1979): *Theoretische Geographie. Grundzüge einer theoretischen Kulturgeographie.* Stuttgart: Teubner(Teubner Studienbücher Geographie).

9 The urban revolution

How thinking about the city in 1920s German geography prepared the field for thinking about quantification and theory

Boris Michel

Introduction

In a recent paper, Trevor Barnes argued for joining urban mathematical models with critical urban social theory. That "the two", he writes, "were separated historically was not because of any philosophical incommensurable difference but a result of internal sociological factors within the discipline of geography during the early 1970s" (Barnes 2019, 491). Barnes begins his paper by stating that he "received a schizophrenic education as it related to the urban" (Barnes 2019, 491), as he experienced it being one half quantitative methods and models and one half critical Marxist theory. "Missing from my undergraduate curriculum, however, was how these two intellectual bodies of work related to one another. It appeared as if they lived in two separate solitudes" (Barnes 2019, 492).

What this brought to my mind was the role of "the urban" and "the city" for geographical thought. Especially with regard to the quantitative revolution in geography. We all know the anecdote told about and by David Harvey about his conversion from being a quantifier to being a radical Marxist as a history of moving from small town England to the burning ghettos of Baltimore in the wake of the assassination of Martin Luther King (Harvey 2000). Eric Sheppard and Barnes recently argued that Baltimore was the "truth spot" for Harvey's work, a place that made his perspective on geography plausible and credible far beyond the city of Baltimore. "Baltimore became a means through which Harvey understood and interpreted Marx" (Sheppard and Barnes 2019, 204). Writing about urban problems and the ghetto formation, Harvey argues that the theoretical models of the quantitative revolution did little to "say anything really meaningful about events as they unfold around us" (Harvey 1972, 6). While urban problems and the city of Baltimore – for Harvey – were the end of the quantitative revolution, the city was also the beginning.

My paper builds on an article I recently wrote with a colleague of mine (Braun and Michel 2019). In the paper we argued that it is worthwhile to read the quantitative revolution through not only new geographical theories, the wider social context of the cold war or new technologies such as the computer,

DOI: 10.4324/9781003122104-9

but through the objects of geographical inquiry (or its "epistemic things" (Rheinberger 1997)). As we know from Thomas Kuhn, "normal science" approaches a crisis when it is no longer able to solve problems or when surprising new problems emerge (Kuhn 1962) – in the social sciences, we would expect those problems to arise from the problematization of social facts and social relations.

The problem that arose – I would argue – was that geographers no longer could ignore the city. But taking "the city" serious would have undermined the dominant paradigm. I will argue that for German quantitative revolution it was exactly the link between a new geographical thought – what would become known as quantitative spatial science – and the city that brought both to the forefront of German geographical research. While the "quantitative revolution" was late to arrive in Germany and the date given in the discipline's historiography is for various reasons 1968/69, I will focus on a prehistory that prepared the ground for the more theoretical and more sweeping attacks on traditional regional geography.[1]

Talking about the quantitative revolution in geography, we are very aware that this was very much focused on social physics and economics and that economic geography played a central role. But we sometimes forget, how this geography was also very urban and part of a turn towards the city. Quantitative geographers left the bucolic and "non-modern" landscapes of Carl Sauer's and others regional geography and moved towards modern urban spaces.

While Sheppard and Barnes argue that Baltimore was a truth spot for Harvey's work and Gieryn who coined the term "truth spot" uses the Chicago of the Chicago School of Sociology's as an example of such a place that gives credibility to delocalized and non-situated scientific claims (Gieryn 2006), I am not interested in one particular city. Instead, I would argue that "the city" as a geographical problem more generally, was some sort of truth spot for quantification, abstraction and theory. The social formation of "the city" helped to make thinking in spatial laws and models more plausible and credible.

Geography and the city: an uneasy relationship

Today, urban geography is one of the central pillars of our discipline. This was not the case in early 20th-century regional geography in Germany (and to my knowledge elsewhere). While one could argue that sociology came to life as an academic discipline because of the city (or rather because of the process of urbanization) and that writers like Max Weber, Ferdinand Tönnies, and Georg Simmel wrote under the impression of rapid urbanization and a new experience of urbanism, geography remained sceptical or disinterested. Whereas "the city" for those sociologists stood for modern society writ large (Weber 1920; Tönnies 1887; Simmel 1903), the urban and geography came together surprisingly late. The term "urban geography" or "settlement geography" as it was called earlier in German geography emerged around 1900 but gained traction only in the 1920s.

If one looks at how the city emerged as a topic for German geography one can find an early encounter in the work of Friedich Ratzel. Most notable in the third part of the second volume of his "Anthropogeographie" (Ratzel 1891) and in one article on "the geographical location of big cities" (Ratzel 1903) where he deals with human habitation and the location of settlements on a broad historical and geographical scale. Friedrich von Richthofen taught "General Settlement and Transportation Geography" in 1891 (Richthofen 1908) and in 1895 Alfred Hetter published a short lecture on "the location of human settlements" (Hettner 1895) in the first issue of Geographische Zeitschrift where he highlights that cities are a key part of regional geography. However, these texts remain marginal within the work of their authors and do little more than locating, describing, and classifying cities.

A first attempt to argue for urban or settlement geography as a proper subfield of geography and to relate this field to the wider methodological and theoretical debates within the discipline was Otto Schlüters "Notes on Settlement Geography" (Schlüter 1899). Schlüter, who a few years earlier published a book on settlements in peripheral regions of Prussia (Schlüter 1896), argues that the city may appear as something that has little to do with the natural landscape and more with ideas or economic factors and thereby would be located outside the scope of regional geography. Nonetheless, he argues, it is "something highly geographical" (Schlüter 1899, 68), an important part of the landscape and therefore has to be "described" and "explained" (those where the terms often used) by regional geographers and by a regional perspective.

What all these texts have in common is not only an attempt to formulate a geographical concept of "city" but that they at the same time indicate the problems of such an attempt. It becomes apparent – most clearly and explicitly in Schlüters work – that the city challenges one of the central pillars of geography, namely the integration of nature and culture. While these authors often highlight that cities were established in a certain location because of the physical geography, their development and history are told as an increasing emancipation from the natural landscape. Nature disappears in favour of a determination by social and economic factors. Thus, the city is ambivalent and somewhat uneasy for a geography that is based on "Länderkunde". And therefore until the 1920s, it is all but certain that the city is a relevant topic for geography that a city is indeed "something highly geographical" (Schlüter 1899). It is open for debate whether a city can be the object of a geographical study and whether one can talk of something like an "urban landscape" (Passarge 1921). Or, on the other hand, whether it has to be regarded as a "small" part of a study on a larger landscape – e.g. the city of Munich taking up two pages in a two-volume 700-page study on Southern Germany (Gradmann 1931).

Overall, early 20th-century German geography is more interested in rural places than in the industrialized and urbanized city. The "Großstadt" – a term that in Germany formally refers to cities with more than 100,000 inhabitants and is often associated with urbanism as a way of life[2] – is problematic, because since the 19th century it figures as the spatial embodiment of modernity and modern capitalism. The urban is associated with acceleration and spatial disembedding. Capitalism, as Marx and Engels wrote a few decades earlier in the

"Communist Manifesto" (Marx and Engels 1972), melts the solid and overturns spatial barriers. The city destabilizes the connection between nature and culture as well as the identity of a spatial individual. Not only are they too social, they are also increasingly determined by factors outside a bounded region. Large parts of geography reacted to this with disinterest or a conservative anti-urbanism. After the First World War, this is especially true for the emerging field of geopolitics, where the city is regarded as a threat to nation, state, and Volk, sometimes with an anti-Semitic dimension, as the city is seen as a place of the cosmopolitan, international, and "rootless" Jewry (Michel 2018).

Urban geography: a new field of geographical inquiry

Nonetheless, in the 1920s, urban geography emerges as a field within German-speaking geography. After the war and due to lost funding, lost colonies, and lost freedom of movement, much of German geography turns more inwards and towards research on Germany, a space that was transformed by rapid urbanization, industrialization, and the war (Wardenga 2001). While one of the consequences of this changed social and political context was the rise of a revanchist political geography or geopolitics, it was also the context that saw an increasing number geographical monographs and PhD thesis emerge that deal with the urban geography of cities (Michel 2016).

In most cases, these studies follow the strict structure, given by Hettners "Länderkundliches Schema" (Hettner 1935), that – at least in the popularized version – calls for a description of landscapes that start with the physical geography, the soil, the climate and then moves upwards to the sphere of human activity or rather to the materialization of human activities in space. This perspective is often historical and morphological and it heavily focuses on visual epistemologies (Michel 2015). Thus, the bulk of this research remains steadily within the well-established framework of classical regional geography, focusing on geographical and topographical locations, historical development and the description of the "bodily appearance of the city" (Carlberg 1926, 153).

What is striking is not only that this remains within the old paradigm but also that this urban geography differs in two more ways from today's urban geography. First, this early empirical urban geography is not interested in large cities and metropoles but in small towns and cities. Today's bias in urban geography for the "non-ordinary" cities, for the global and mega cities and for a small number of well-researched iconic cities, is absent. There is hardly any study on the large urban centres of modern Germany and there is hardly more than one study on any city. The first urban geography on Berlin is published in 1933 (Leyden 1933). Instead, most work is done on cities with less than 50,000 inhabitants. These are cities that are perceived as being "rooted" in their places, that are only slowly changing, that are not "unnatural". Second, this urban geography is not interested in differences within cities. Difference is difference between spaces, not within spaces. So we find differences between "the Oriental city" and the "Southern German city", between the city of Kiel and the city of Rostock, but no divided cities, not much on spatial segmentation within a given city, no class division, and no social conflicts. Nonetheless urban geography is established as a field within the discipline and in 1935, reviewing Walter Christallers "Die zentralen Orte in Süddeutschland" (Christaller 1933), Hans Bobek writes that the time has long gone, when "cities in a way where perceived as foreign objects" in regional geography (Bobek 1935, 125).

But while this first boom is clearly rooted in old and conventional geography, in the late 1920s this becomes the empirical field where we can find a "modernization" of geography – both, in terms of something new and in terms of a turn to modernity. What was important for this – and to new thinking in geography in many cases – was a mobilization of people and ideas outside of geography. These new allies, as one could call them in Latourian terminology were often found in economics and sociology – though German geographers at that time are really bad at citing references. One of the key protagonists is Bobek, who published his dissertation on the Austrian city of Innsbruck in 1928 (Bobek 1928), that was accompanied by a theoretical paper on the fundamentals of urban geography (Bobek 1927) that was often regarded as the first contribution to a modern German urban geography (Schöller 1969).

Bobek criticized urban geography for its focus on cities as part of a cultural landscape or as cultural landscapes in their own right, its focus on morphology and historical development. More interesting, he argues, is the city as "a living economic entity" within the wider economics of the given landscape (Bobek 1927, 214). While "Länderkunde" was challenged by a number of authors in the late 1920s and others highlighted the role of dynamic spatial forces ("Kräfte") for understanding geographical regions, Bobek is today seen as the founding father of a functionalist geography. Instead of the urban morphology, it is the function of urban elements that become central for Bobek. Not the city as an entity or a thing, but its relations to the wider region and other cities are what this new urban geography has to focus on. Be those relations economic, political or social. Here, Bobek is heavily influenced by the work of German sociologist Werner Sombart's work on modern capitalism and his urban theory (Sombart 1916, 124–133). What makes a city, for Sombart, is less its build form, and not those people and functions that "fill" the city, but those that "generate" the city, i.e. give it a function in relation to a space beyond its own. In this emerging functionalist urban geography, we also start to see functional and social difference within a city.

There are two moments, which are important for my argument. First, this urban geography distances itself from the holism of regional geography. Even monographs now tend to focus on more than just one certain spatial unit that is the object of a totalizing description. Second, this geography distances itself from the focus on the material, static and often visual elements in space. Both moments indicate a growing interest in abstraction and abstract objects as well as methods of geographical inquiry.

The best known example of such a perspective is without a doubt Walter Christallers "Theorie der zentralen Orte" (Christaller 1933), a book that abstracts and reduces the geography of Southern Germany to simple functions in a way that might appear scandalous to a traditional geographer. Searching spatial laws in the distribution of settlements through an analysis of the distribution of telephone connections and businesses was not something traditional geography would think of as geography.

The case of Christaller is interesting for my argument for another reason. While we tend to read Christaller's work today as "theoretical geography", this decouples him from his connections to urban geography. Not only was his dissertation supervised by Robert Gradmann who was amongst the earlier urban geographers (Gradmann 1914) and was the aforementioned review by Bobek (Bobek 1935) framing the book as a work of urban geography. This framing was ambivalent and changed. A telling anecdote is published in one of the 1970s textbooks for theoretical geography, written by Eugen Wirth, professor at the same institute where Christaller submitted his dissertation and who told his own story about the quantitative revolution (Paulus 2017). In his introduction to theoretical geography, he writes how over the last 40 years Christaller's book in the library of his institute was moved from regional geography of Southern Germany to urban geography and finally to theoretical geography (Wirth 1979).

Urban Social Geography after 1945

It is this functionalist urban geography that prepares the ground for what from the early post-war period onwards is called "Sozialgeographie", social geography. It again is Hans Bobek who, in the second volume of the first journal to be published after the war, coins the term as a key term for a new modern human geography (Bobek 1948). What traditional German geography was lacking, he argues, was a concept of "society" and "the social". Its agents until now were states, "people" and "man". If geography wants to have a place in German academia after National Socialism, it needs to build a bridge to sociology and apply a concept of society that is functional differentiated. Not "man" creates cultural landscapes, but different social groups shape it in different ways. "Social geography" is one of the key fields and terms of the quantitative revolution in the late 1960s. It signifies the new modern geography, both in terms of methods and empirical topics, even though "Länderkunde" remains one of its foundational concepts. Bobek highlights on his several occasions (Bobek 1948, 1962, 157) that this social geography was built on empirical and theoretical challenges in the emerging field of urban geography.

Urban geography after 1945 continues its functionalist agenda and we see a further growth of the field within the discipline. A few points about urban geography in post-war West Germany are worth mentioning:

1 Urban geography is increasingly interested in larger cities. Those cities are frequently addressed as "modern cities" and those modern cities are functional differentiated – especially with regard to their economic sectors and neighbourhoods. Thus, something like a monograph on the social

geography of one shopping street in downtown Frankfurt becomes something that qualifies as geography (e.g. Hübschmann 1952). The anti-modernism that characterized early urban geography and "Länderkunde" in general is fading.

2 This urban geography is also increasingly interested in current problems and not only in description. Those problems at that time were a drastic urban crisis, post-war reconstruction, and the emergence of modernist urban and spatial planning. Thus we see a geography that is increasingly interested in being an applied science (Wardenga et al. 2011). Thereby, the city is increasingly becoming a term that addresses a specific modern form – whereas Ratzel used it in an explicitly ahistorical way (Ratzel 1903). This becomes especially prominent in discussions around reconstruction. While many inner cities were rebuilt in a traditionalist architecture, we also start to see arguments made that "The medieval old-towns with their narrow building, and tight and bent lanes are an obstacle for the free development of European metropoles". Thus the "gruel experiment, aerial warfare conducted on the old-towns" (Kraus 1953) opens up opportunities for modern urban planning and development.

3 While far from being a theoretical spatial science in the sense of the Anglophone geography of the same time and while descriptive "Länderkunde" remains a central pillar of geography writ large and of urban geography, this urban geography, "theory" and "concepts" play an increasing role. As "the modern city is the most complicated chain in our highly organized cultural landscape" (Schöller 1973) it also becomes clear that urban geography and a geography of cities call for new methodological approaches that go beyond what regional geography did. Be it that thematic mapping or interviews enter the canon of geographical methods (Schöller 1953) or that someone like Christaller, "telephone method" condenses the "meaning of a place" in one "common denominator" (Christaller 1933, 142) in order to make an abstract concept visual in the form of hexagons.

These three points, 1) an interest in a different form of urban spaces and cities, 2) an interest in issues of modern urban development and planning and 3) new modes of geographical research and terminology, indicate a modernization of urban geography without a challenge to "Länderkunde" in general. The city is a special case of Länderkunde.

Conclusion

In this paper, I argued that quantitative thinking in geography becomes plausible only after geography starts to become interested in a new geographical space, the modern city. It should also have become clear that this modern city could hardly be addressed by classical regional geography.

To conclude: Why is the city challenging regional geography? On the one hand – as I pointed out – the modern city was seen as a threat to a discipline that is held together by the integration of nature and culture. Thereby, geographers often looked at cities with a considerable unease. The anti-modernism of post-World War I geography pushed geography further to an anti-urban sentiment. Thus, regional geography was increasingly incapable of talking about spaces that increasingly dominated the experience of geographers and that became especially important as a political topic after 1945.

However, I think there is more. Cities as the spaces of modernity and modern capitalism are also a more fundamental challenge for a regional geography. And I will conclude by some simple dualisms: While regional geography's temporality was the slow change of natural and cultural landscapes that took place over centuries, the city since the late 19th century is speed and acceleration and after 1945 cities were often rebuilt by abstract plans from scratch without respect for the physical geography. While landscapes are unique and understood as "geographical individuals" (Ritter), modern capitalism and modern urban planning are seen as universalizing forces that glosses over local differences by turning everything into exchangeable commodities. While landscapes are concrete and landscapes are things, the dynamics governing modern capitalist societies are abstract relations. While regions are bounded spaces, the modern city is not without the world market.

Moreover, we can find this argument in the dissertation of Dietrich Bartels, who in the late 1960s becomes the most prominent author of the quantitative revolution in Germany (Bartels 1968). I think it is no coincidence that he wrote his dissertation on twin cities in the style of functionalist urban geography (Bartels 1960). Bartels argues not only that functionalist geography is especially important for the study of cities but also that the urban phenomenon and flows are increasingly separated from morphology and a determination by physiognomy. The social formation of the city can no longer be "extracted from the image ["Strukturabbild"]" and city's functions are not neatly bounded. Thus, the urban dynamics and the permanence of the build form contradict each other and, as Bartels argues, the study of urban geography has to go from the "study of the state of things to the processes, from the external image if cities to the decisive inner life" (Bartels 1960, 1–3).

In Marxist terminology, one could argue that regional geography is not a paradigm for modern industrial capitalism. As German geographer and 1980s lone Marxist Ulrich Eisel puts it:

> The [traditional] geographical paradigm, its structure and the time of its origin, is only comprehensible if one understands it as a bulwark against all modern processes of abstraction. How this is constructed on the level of spatial theory has been explained [in the book]. The material core of such processes of abstraction was located in the industrial capitalist mode of production. Urbanization, democracy, empirical science, etc. go hand in hand with it. These are all forms of destruction of traditional orders. The ideal of

> such orders is determined by the so-called idiographical world view that the geographers insisted on. In this respect, I described geography as a kind of anti-industrial reaction to industrial progress. This reasoning resulted from the paradigm's core: it was a euphoric plea for concrete living conditions. So it had to be directed against the universal power of abstraction of the industrial mode of production.
>
> (Eisel 2008)

This was not a problem as long as geographers were primarily exploring the "non-urban" frontiers of capitalism, be it in the context of colonial geography, the internal hinterlands or historical geographies. But once geography turned towards the territory of the capitalist core, this was ridiculously out of touch with the lived experience and with the expectations of a modern science.

Notes

1 I use the term "regional geography" as a shorthand for the dominant paradigm of German geography between the late 19th and mid-20th century, a geography that was addressed as "Länderkunde" or "Landschaftsgeographie" and that is most prominently linked to the work of Alfred Hettner (Hettner 1927). The most direct translation to Anglophone geography can be found in the work of Carl Sauer (Sauer 1925).

2 Thus, Simmel uses the term in his paper "Die Großstädte und das Geistesleben", which has been translated as "The Metropolis and mental life".

References

Barnes, T. (2019): Not Only . . . But Also: Urban Mathematical Models and Urban Social Theory. In D'Acci, L. (Ed.): *The Mathematics of Urban Morphology*. Cham, Switzerland: Birkhäuser (Modeling and Simulation in Science, Engineering and Technology), 491–497.

Bartels, D. (1960): *Nachbarstädte. Eine Siedlungsgeographische Studie anhand ausgewählter Beispiele aus dem westlichen Deutschland*. Bad Godesberg: Bundesanstalt für Landeskunde und Raumforschung.

Bartels, D. (1968): *Zur wissenschaftstheoretischen Grundlegung einer Geographie des Menschen*. Wiesbaden: Franz Steiner.

Bobek, H. (1927): Grundfragen der Stadtgeographie. In *Geographischer Anzeiger*, 28, 213–224.

Bobek, H. (1928): *Innsbruck. Eine Gebirgsstadt, ihr Lebensraum und ihre Erscheinung*. Stuttgart: J. Engelhorn.

Bobek, H. (1935): Eine Neue Arbeit zur Stadtgeographie: Rezension von Walter Christaller, Die Zentralen Orte in Süddeutschland. In *Zeitschrift der Gesellschaft für Erdkunde zu Berlin*, 125–130.

Bobek, H. (1948): Stellung und Bedeutung der Sozialgeographie. In *Erdkunde*, 2, 118–125.

Bobek, H. (1962): Kann die Sozialgeographie in der Wirtschaftsgeographie aufgehen? In *Erdkunde*, 16, 119–126.

Braun, J.; Michel, B. (2019): "Das komplizierteste Glied unserer hochorganisierten Kulturlandschaft". Die Anfänge der quantitativ-theoretischen Wende und das Problem der Stadt. In *Geographische Zeitschrift*, 107(2), 88–106.

Carlberg, B. (1926): Stadtgeographie. In *Geographischer Anzeiger*, 27, 148–153.

Christaller, W. (1933): *Die Zentralen Orte in Süddeutschland*. Jena: Gustav Fischer Verlag.

Eisel, U. (2008): Moderne Geographie mit atavistischen Methoden. Über die undeutliche Wahrnehmung eines deutlichen Paradigmas. In Eisel, U.; Schultz, H. (Eds.): *Klassische Geographie. Geschlossenes Paradigma oder variabler Denkstil? Eine Kritik von Ulrich Eisel und eine Replik von Hans-Dietrich Schultz*. Berlin: Geographisches Institut der Humboldt-Universität zu Berlin, 1–37.

Gieryn, T. (2006): City as Truth-Spot: Laboratories and Field-Sites in Urban Studies. In *Social Studies of Science*, 36(1), 5–38.

Gradmann, R. (1914): *Siedlungsgeographie des Königreichs Württemberg*. Stuttgart: Verlag von J. Engelhorns Nachfahren (Forschungen zur deutschen Landes- und Volkskunde, 21(2)).

Gradmann, R. (1931): *Süddeuschland*. Stuttgart: Engelhorn, 2 volumes.

Harvey, D. (1972): Revolutionary and Counter Revolutionary Theory in Geography and the Problem of Ghetto Formation. In *Antipode*, 4(2), 1–13.

Harvey, D. (2000): Reinventing Geography. In *New Left Review, Second Series*, 4, 75–97.

Hettner, A. (1895): Die Lage der menschlichen Ansiedelungen. Ein Vortrag. In *Geographische Zeitschrift*, 1(7), 361–375.

Hettner, A. (1927): *Die Geographie*. Breslau: Ferdinand Hirt.

Hettner, A. (1935): Das Länderkundliche Schema. In *Geographischer Anzeiger*, 1–6.

Hübschmann, E. (1952): *Die Zeil. Sozialgeographische Studie über eine Straße*. Frankfurt am Main: Kramer.

Kraus, T. (1953): Die Altstadtbereiche Westdeutscher Großstädte und ihr Wiederaufleben nach der Kriegszerstörung. In *Erdkunde*, 7, 94–99.

Kuhn, T. S. (1962): *The Structure of Scientific Revolutions*. Chicago: University of Chicago Press.

Leyden, F. (1933): Berlin als Beispiel einer wurzellosen Großstadt. In *Zeitschrift für Geopolitik*, 10(3), 176–188.

Marx, K.; Engels, F. (1972): Manifest der Kommunistischen Partei. In Marx, K.; Engels, F. (Eds.): *Werke*. Berlin: Dietz, volume 4, 459–493.

Michel, B. (2015): Geographische Visualitätsregime Zwischen Länderkunde und quantitativer Revolution. In Schlottmann, A.; Miggelbrink, J. (Eds.): *Visuelle Geographien*. Bielefeld: Transcript, 209–224.

Michel, B. (2016): "Man sieht es und hört es und fühlt es, dass man in einer ungeheuren Maschine Steckt, in der Seltsamsten, Welche je die Menschen Erfunden Haben". Zur Geschichte der Stadtgeographie vor 1945 und zur Frage von Geographie und Antimodernismus. In *Berichte. Geographie und Landeskunde*, 90(1), 5–24.

Michel, B. (2018): Anti-Semitism in Early 20th Century German Geography: From a "Spaceless" People to the Root of the "Ills" of Urbanization. In *Political Geography*, 65, 1–7. DOI: 10.1016/j.polgeo.2018.03.006.

Passarge, S. (1921): *Vergleichende Landschaftskunde*. Berlin: Dietrich Reimer.

Paulus, K. (2017): Revolution Ohne Kiel und Ohne Revolution – Die Quantitativ-Theoretische Geographie in Erlangen. In *Geographica Helvetica*, 72(4), 393–404. DOI: 10.5194/gh-72-393-2017.

Ratzel, F. (1891): *Anthropogeographie. Zweiter Teil. Die geographische Verteilung des Menschen*. Stuttgart: J. Engelhorn.

Ratzel, F. (1903): Die Geographische Lage der großen Städte. In der Gehe-Stiftung, J. (Ed.): *Die Großstadt. Vorträge und Aufsätze zur Städteausstellung*. Dresden: Jann & Jaentsch, 33–72.

Rheinberger, H. (1997): *Towards a History of Epistemic Things: Synthesizing Proteins in the Test Tube*. Stanford: Stanford University Press.

Richthofen, F. V. (1908): *Vorlesungen über allgemeine Siedlungs- und Verkehrsgeographie. Bearbeitet und herausgegeben von Dr. Otto Schlüter*. Berlin: Dietrich Reimer.

Sauer, C. (1925): The Morphology of Landscape. In *University of California Publications in Geography*, 2(2), 19–54.
Schlüter, O. (1896): *Siedlungskunde des Thales der Unstrut von der Sachenburger Pforte bis zur Mündung*. Halle: Wischan & Wettengel.
Schlüter, O. (1899): Bemerkungen zur Siedlungsgeographie. In *Geographische Zeitschrift*, 5, 65–84.
Schöller, P. (1953): Stadtgeographische Probleme des geteilten Berlin. In *Erdkunde*, 7(1), 1–11.
Schöller, P. (Ed.) (1969): *Allgemeine Stadtgeographie*. Darmstadt: Wissenschaftliche Buchgesellschaft.
Schöller, P. (1973): Tendenzen der stadtgeographischen Forschung in der Bundesrepublik Deutschland. In *Erdkunde*, 27(1), 26–34.
Sheppard, E.; Barnes, T. J. (2019): Baltimore as Truth Spot. In Barnes, T. J.; Sheppard, E. S. (Eds.): *Spatial Histories of Radical Geography: North America and beyond*. Hoboken, NJ: John Wiley & Sons Ltd (Antipode book series), volume 6, 183–209.
Simmel, G. (1903): Die Großstädte und das Geistesleben. In der Gehe-Stiftung, J. (Ed.): *Die Großstadt. Vorträge und Aufsätze zur Städteausstellung*. Dresden: Jann & Jaentsch, 185–206.
Sombart, W. (1916): *Der moderne Kapitalismus*. Zweite, neugearbeitete Auflage. Erster Band. München: Duncker & Humblot.
Tönnies, F. (1887): *Gemeinschaft und Gesellschaft*. Leipzig: Fues.
Wardenga, U. (2001): Theorie und Praxis der länderkundlichen Forschung und Darstellung in Deutschland. In Grimm, F.; Wardenga, U. (Eds.): *Zur Entwicklung des Länderkundlichen Ansatzes*. Leipzig: Inst. für Länderkunde (Beiträge zur regionalen Geographie, 53), 9–35.
Wardenga, U.; Henniges, N.; Brogiato, H. P.; Schelhaas, B. (2011): *Der Verband Deutscher Berufsgeographen 1950–1979. Eine sozialgeschichtliche Studie zur Frühphase des DVAG*. Leipzig: Leibniz-Inst. für Länderkunde (Forum IfL, 16).
Weber, M. (1920): Die Stadt. Eine soziologische Untersuchung. In *Archiv für Sozialwissenschaft und Sozialpolitik*, 47, 621–772.
Wirth, E. (1979): *Theoretische Geographie. Grundzüge einer theoretischen Kulturgeographie*. Stuttgart: Teubner (Teubner Studienbücher Geographie).

10 A revolution in process

Longue Durée and the social history of the increase in numerical data from the Brazilian Institute of Geography and Statistics and the National Geography Council before the "quantitative revolution" (1938–1960)

Larissa Alves de Lira[1]

Abstract

The concept of "Revolution" does not fit well with the idea of a slow historical process. In the domain of the history of geography, historians often talk about a "quantitative revolution" that has emerged in many countries since 1960. Is this how we must interpret Brazilian Geography? The Brazilian Institute of Geography and Statistics (IBGE) was founded in 1938 and the modernization of statistics in Brazil began in 1940. From 1940 to 1960, did Brazil accumulate enough series of numbers to be considered part of the movement of a "quantitative revolution"? This paper aims to analyze how numbers "advanced" in the IBGE's production of images between 1938 and 1960, and to evaluate the success of data accumulation and the advancement of the mathematical point of view in geography just before the so-called "quantitative revolution."

Introduction

The aim of this article is to evaluate the increase in the numerical data in the works carried out by geographers associated to the Brazilian Institute of Geography and Statistics (IBGE) and to the National Geography Council (CNG) published in the *Revista Brasileira de Geografia* between 1938 and 1960 in Brazil. I understand this evolution as the process of accumulating numerical data on Brazil over space and time. I conducted this evaluation for the period before the emergence of the so-called "quantitative geography" around the world and in Brazil, considering 1960 as the date attributed to the beginning of a "revolution" of practices and theoretical conceptions of the discipline.[2] This evaluation

DOI: 10.4324/9781003122104-10

is, more specifically, about constructing a hypothesis for the debate, which I present in the following.

Firstly, the national Brazilian State began to associate geography with statistics long after Europe did. From the moment statistical expertise was incorporated however, the country implemented it relatively fast, from an exclusively statistical point of view.

This intellectual movement took place in order to support the process of territory modernization[3] with infrastructures capable of placing the country on the list of the great modern nations. Nevertheless, despite the fast advancement of statistics themselves, the *social process* as a whole is incomplete, not as much from the perspective of statistical data accumulation as from the projects and aspirations they supported.

Knowing the country's morphologic *measurements*, in its physical and human dimensions, was essential to bringing a more pragmatic dimension to the plans for modernization. Intellectuals aimed to leave behind the heritage of slavery (Machado 2012) and create a new image for Brazil at the turn of the 1930s (Lira 2021). However, despite the desire to conduct broad mathematical diagnostics and the modernization plans that would accompany them, a slower rhythm set in regarding the measurements and the accomplishment of *social projects* that this intellectual movement would support. This was due to the fact that the movement faced some *longue durée* (from here on long-term), social, and also physical embarrassments.

Both the speed and the quality of this quantitative advancement were affected by the constraints brought by social and spatial structures. The social and political structure inherited from an oligarchic and slavery-based society halted these plans despite the relative speed with which statistics themselves were gathered in Brazil. I argue that the reasons for this general process having been hampered must be analyzed from a social history perspective. In addition, the plans, more so than the mathematical diagnostics, were hindered by the size of the country. There was not enough private and public capital to finance them due to Brazil's size.

I also plan to evaluate statistics and their expansion based on the history of science considering spaces and their long-term temporality; a perspective known as geohistory of knowledge, associated with social history, as I will explain later. In the long term, statistics were associated with a utilitarian mindset developed after the Enlightenment and the Industrial Revolution (Quaini 1983).

For that matter, I believe that Brazilian geography developed *beyond* a movement of at least two long durations of scientific mentalities. Firstly, a tension existed between literary and utilitarian values of geography. These mindsets contradict each other, but the utilitarian values almost always win this game in a context of rapid development of techniques and of capitalism. Literary values, however, represent, in the long run, the influence of romanticism on geography, constantly replacing their postulates. The final moment of this tension, which represents the victory of one of the sides of this scale, was not concluded until at least 1978 in Brazil. This is the same as saying that, until this date, Brazilian geography was less literary than its French counterpart, but less pragmatic than the geography produced in the United States, its other source of inspiration. Still, it incorporated both sides of this tension. Secondly, in Brazil, the drive for development had already been in place for some time (the term "developmentalism" would come later), meaning development was seen as urgent and feasible for the country. This state of seeking the latent development circling the country at the time also configured a mindset, an infra-epistemology of geography specific to Brazil.

In this movement, Pierre Deffontaines,[4] Pierre Monbeig,[5] and Francis Ruellan[6] had undeniable influence on the first years of development of Brazilian institutional geography as well as on the global scientific mindset that emerged in the country. A monographic and descriptive geography was broadly practiced in Brazil before 1960 in parallel with the inexorable advance of statistics. Later, the IBGE and the CNG, in broadening their diplomatic policy, would receive geographers from other nationalities, including Leo Waibel (1888–1951), a German naturalized American, and American Preston James (1899–1986), both practitioners of theoretical frameworks more open to statistics. In this context, a kind of epistemological dispute took place between these characters or what they represented in terms of geography.[7]

Methodology

As previously stated, my methodology is that of social history, including a perspective of the geohistory of knowledge. I understand the social history of science as the possibility of thinking of scientific fact as socially constructed. In this sense, science is mainly involved in a social, cultural, and political context. The dimensions of society are interconnected, and scientists, politicians, and other actors are involved in parallel negotiations and plans surrounding the legitimate use of science.

On one hand, I observe scientific phenomena developing in broad frameworks of time and space, inspired by the methodology created by the French historian Fernand Braudel (Braudel 1992), while supposing the existence of temporalities and infra-mindsets – deep movements that take place in rhythms different to superficial ones. On the other hand, this methodology aims to place Brazil in the sphere of a "world-science" (Polanco 1989), in the sense of a global scientific community. Alongside this geohistory, I attempted to trace a broad contextualization of the social history of the statistical movement related to geography as it took place in Brazil. Writing a history of statistics is beyond the scope of the project which will focus on tracing a structural movement of statistics connected to geography.

I have elaborated my own methodology for verifying the advancement of the statistical movement associated to geography and produced graphs showing an evolution of the use of numbers through time. The documents that compose my corpus for the quantitative analysis are articles published by the *Revista Brasileira de Geografia* between 1939 and 1955. The *Revista Brasileira de Geografia* was a joint publication by the IBGE and the CNG founded in 1939. My first goal was to list the ways statistical data could be presented in text by counting how many tables and images (graphs and maps) with scales and geo-references were published per year. Tables, images, and maps were considered as units of measurement.

In total, I analyzed 1245 publications (including all papers, research notes, and news reports) in 65 journal issues. In these publications, there were 1395 figures (tables, images, and maps) which corresponded to my criteria. I then separated this phenomenal form of numbers in the text according to the following criteria: whether they were related to a degree of national coherence (related to Brazilian space as a whole), to a regional degree (above or below the scale of states),[8] to a state degree (on the scale of states), or to a municipal/urban degree (on the scale of municipalities and cities). In other words, I evaluated on which scale and in which referential space of analysis the numbers were mobilized. For this evaluation, I analyzed the images themselves (in the case of maps) or the titles of the tables.[9]

Geo-referenced, scaled images (showing the location of statistical analysis), tables, and maps were understood as expressions of a process of mathematization of the territory connected to geography. Locating a phenomenon by giving it dimension, even if that meant only a vague approximation of latitude and longitude, was understood as a process of geographical understanding based on mathematical thinking. Almost all occurrences of images and numeric tables were considered.

Photos, drawings, profiles with no numeric occurrence, blocks, diagrams, and sketches were not considered. I understand these phenomenal forms as expressions of observations and descriptions of landscapes. Analyses of samples, such as chemical analyses of rocks, were disregarded due to a lack of location. Theoretical, doctrinal, and didactic articles or regionalization proposals, even those containing numbers and tables, were also disregarded, since I understand they would not add new numbers to the territory analysis. The quantitative analysis was carried out up until the year 1955. Despite publications at later dates, the number of pages and articles in the journal decreased greatly after 1955, which could conceal the evolution of the original corpus.

Regarding the maps, I conducted a qualitative analysis ranging from 1939 to 1960, showing the evolution of their forms and lines. To do so, I used just one "law" of Graphic Semiology (Bertin 1967) as a reference. According to the development of this theory, finer lines with more precise curves represent more accurate data. Thicker lines represent less detailed data. Finally, I selected all articles elaborating plans for the country, projects, and territorial planning, both urban and regional, published between 1939 and 1960 and which expressed the developmentalist plans associated with this statistical diagnosis.

Cartographic, developmentalist, and statistical predecessors in Brazil

Brazil faced major problems at the beginning of 1930, just after the Brazilian Revolution of 1930. But even before this, the country had undertaken several developmentalist missions.[10] Brazil's problem was seen, after the imperial era (1822–1889) and the "Old Republic" (1890–1930), as the dissemination throughout its territory of starving, barefoot, sick people. Thus, an "applied" human geography was tasked with the process of including Brazil in the list of modern societies and territories. Candido Rondon, a marshal in the Brazilian army, undertook a mission to populate the territory with telegraphic lines during 1907 while studying the most remote areas in Brazil, concluding that the country had abundant resources (Maciel 1998).

But what were the dimensions of these resources? Carlos Augusto Figueiredo Monteiro (2002) considered the period between 1900 and 1935 that which preceded scientific geography in Brazil. The author believes that at the beginning of the 20th century a reasonable amount of knowledge had been accumulated in geographical terms. The more important states in the Brazilian federation, such as São Paulo and Minas Gerais, already had permanent commissions for Cartography, Geography, and Geology. In 1868, the Brazilian Empire published an atlas with a map of Brazil divided into its regional dimensions. On the general map as well as on the regional maps, there is a clear geodesic effort related to locating toponymies, forms of relief, and rivers (Figure 10.1).

At the beginning of the 1920s, when the country had already become a Republic, right after the Rondon Commission efforts, the Engineering Club, under the leadership of Francisco Behring, also founded a commission to enhance cartographic techniques and Brazilian maps. This commission aimed to include Brazil in the project of the world's millionth map initiated by Oscar Peschel in Germany (Duarte 2013). Topographic studies also involved deep mathematical knowledge, instruments, and calculations. This commission also published a new map of Brazil (Figure 10.2).

In Bahia, using other dimensions and scales of analysis, urban improvements were being made by Teodoro Sampaio, including actions concerned with promoting hygiene and vaccinations (Sousa and Vaz 2019). The 1920s are also marked by an economics-based perspective applied and committed to the sphere of education. One of the people responsible for modernizing geographic education in the country was Carlos Delgado de Carvalho (1884–1980), who wrote about subjects ranging from physical geography to the Brazilian economy (Machado 2000).

The formation of the National State and planning

The Brazilian National State's modern period started with the Brazilian Revolution of 1930 led by Getúlio Vargas. Vargas' rise to power, as well as that of the military forces from Rio Grande do Sul, ended with the old oligarchic hegemony that dominated the earlier federal government, with fluctuations between politicians from São Paulo and Minas Gerais, which had important consequences for Brazil. The revolution led to a rebalancing of forces in the national federative pact and culminated in conflict between the new federals and the inhabitants of São Paulo (*Paulistas*) in 1932.

Vargas intended to create a State based on economic planning and intense public investment in infrastructure and base industries. The National Institute

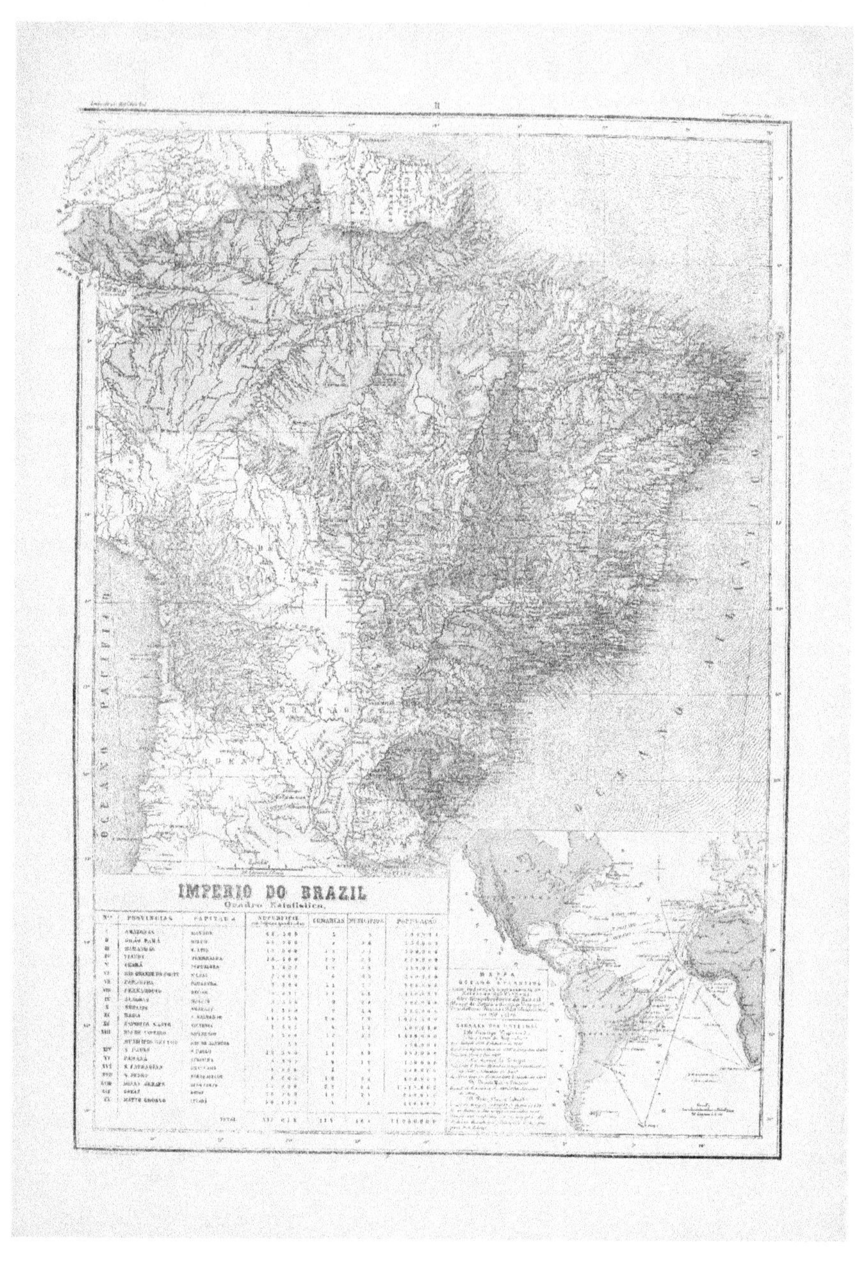

Figure 10.1 Imperio do Brazil (The Empire of Brazil).
Source: Almeida (1868, 39)

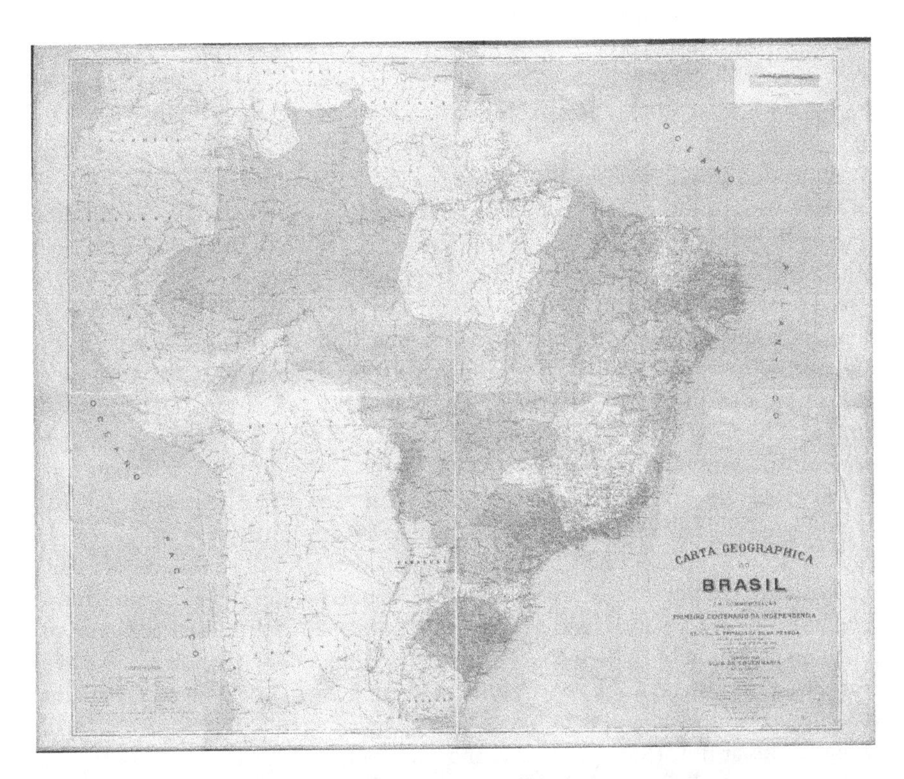

Figure 10.2 *Carta Geographica do Brasil*, 1922 (Geographical map of Brazil). Scale: 1: 100 000.

Source: http://portalclubedeengenharia.org.br/2019/01/11/o-clube-de-engenharia-na-confeccao-da-carta-geografica-do-brasil-para-a-comemoracao-do-1o-centenario-de-independencia-do-brasil/

of Statistics (INE) was created in 1936 to conduct systematic surveys at the federal, state, and municipal levels; federal government was made responsible for organizing Brazilian statistics (Senra 2014). An attempt was also made to insert Brazil into the international statistics standards (Senra 2014).

The government created a provisional Executive Board, choosing Teixeira de Freitas as the Secretary General (Senra 2014). Teixeira de Freitas later returned to the "state of the art" of Brazilian statistics previously employed which "brought discontinuity, not allowing for a secure basis for the studies necessary to carry out solid administrative order and achieve progress in the country" (Teixeira de Freitas apud Senra 2014). At the same time, a Brazilian Council of Geography was created (Senra 2014). "The new Council was soon after created by Decree 1,527 of 24 March 1937" (Senra 2014) and would be incorporated into the INE (Senra 2014). All Brazilian statistics and cartography would be reorganized favoring municipal and national scales (Santos and Castiglione 2014).

Hired by the state and federal governments, Pierre Deffontaines, Pierre Monbeig, and Francis Ruellan were, as foreign geographers, the main participants of the French missions that came to Brazil to fund the first modern universities in the country. Pierre Monbeig made his home in São Paulo. In 1935, after a short stay in São Paulo, Pierre Deffontaines went to Rio de Janeiro and Francis Ruellan joined him. The two geographers came to Brazil inspired by the regional, monographic tradition of French geography, but their actions in the country went beyond stimulating descriptive, regional monographs.

The participation of the French geographers was also important to encourage the founding of the IBGE, combining the expertise of the CNG with that of the INE. This shows that the efforts to stimulate data collection were also a global tendency. In this way, the CNG counted on the support of Pierre Monbeig and other even more notable Frenchmen. Emmanuel de Martonne, for instance, an undeniable master of French geography, would have conditioned the creation of the CNG to Brazil's entry in the International Geographical Union. The CNG was also relevant for scientists at the time since it was the cause of the fusion of geographic research in the INE. The entity brought the 1940 census into force in 1938. However, there was great dispute within this institutional environment regarding the use of statistics.

Therefore, Brazil became a territory of exchanges with the insertion of many international geographical movements in the country. Thus, I understand Brazil as a fusion of perspectives, and competition between international branches of geography, seen here in their long-term development. The French perspective was very strong. In addition, Von Thünen's theory, after circulating in Germany, the United Kingdom, and the United States, is finally put into practice in Brazil in the 1950s at the IBGE (we will come back to this subject).

According to Elvin Wyly (2019), the proposal of a monographic and descriptive geography, such as a French-inspired modern geography, loses ground after the battle to keep geography as an academic discipline in Harvard in 1945.

In 1945 – in other words, in the middle of this debate – Edward Ackerman, future president of the Association of American Geographers, published the article "Geographic training, Wartime Research, and Immediate professional objectives," in which he states that the geographer must study foreign languages, abandon philosophy, and specialize in techniques. After the dispute with Harvard, Ackerman began working as a consultant for companies and the State on war-related areas. In this vision, statistics should be used to mobilize resources and populations. This debate is pertinent to Brazil.

Controversies in the Brazilian context: an epistemological approach to numbers

From what I have shown, it is clear that after 1935 Brazil was part of this international arena in which a conflict of characters and mindsets took place. Controversies existing in the context of the constitution of the CNG and the IBGE are an expression of mindsets in opposition and reaffirm the different characteristics of these ways of understanding geography and the value of analyzing them in a way that is also epistemological. Certain debates in Brazil showed which epistemologies were in conflict and in turn the problem of statistics, projective thinking, and of the effectiveness of geography in reorganizing the territory.

This conflict would become clear in 1944. CNG's general secretary was organizing the II Pan-American Meeting for Consulting on Geography and Cartography, having been to the US often (Penha 1993, 39). The meeting took place in Rio de Janeiro in August and the geographers from IBGE were highly active (Penha 1993). In the same year, IBGE's geographer Jorge Zarur published work in the *Revista Brasileira de Geografia* advertising the movement for geography renovation in the US and "the fight of these geographers, who labored to remove geography from *purely academic or laboratorial instances* to place it *at the service of man*, making it a basic, *useful tool* for *administrators and planners*" (Penha 1993, emphasis added, [translated by the author]).

Following this, the problem of statistics entered into the debate in Brazil. In 1951, General Poli Coelho became president of the IBGE, replacing the long tenure of José Carlos de Macedo Soares Guimarães (1936–1951). This brought up a serious discussion on the role of statistics, which clearly opposed a historical, monographic, and descriptive construction of knowledge instead of mathematical reasoning. Poli Coelho asked Lourival Câmara to write a report on IBGE's work, which determined that the "IBGE was much more interested in cultural and political matters than in purely statistical ones" (Almeida 2000 [translated by the author]).

Finally, it becomes evident that this theoretical conflict involved reorganization plans that were important for Brazil, such as the matter of colonization and the choice of a new location for the country's capital. There was also conflict within the CNG regarding which city would be the capital. French geographers Francis Ruellan and Leo Waibel were on the foreigners' side (Adas 2006) and Teixeira de Freitas, Poli Coelho, Fábio de Macedo Soares Guimarães, and Jorge Zarur on the national side.

These disagreements are examples of how the opposition between utilitarian, mathematical geography and monographic, historical geography and their poles, United States and France, are also present in the context of epistemology of geography and of territorial planning. However, despite these disputes, the statistical movement gained ground in Brazil.

The advance of statistics by the IBGE and the CNG in the *Revista Brasileira de Geografia*

The tables, images, and maps produced by the IBGE and the CNG and published in the *Revista Brasileira de Geografia* between 1939 and 1955 portrayed the rapid increase in numbers. In 1938, Brazil was preparing its first modern census, without breaks and using reliable data acquisition methods. The census of 1940 was inspired by the work the IBGE's president, Teixeira de Freitas, had already conducted in Minas Gerais.

Considering the period under analysis here, two dates are important from a data accumulation perspective: only in 1947 was the data from the 1940 census translated into images (its results being transformed into maps) and more broadly used in geography papers. The next census was conducted in 1950, this time with faster data incorporation: in 1952 the first results could be seen in image form. It is also important to note that in 1946, with the end of the New State and the beginning of General Dutra's administration, the *Revista Brasileira de Geografia* was much less important, going from having an average of 200 pages per issue to 140 pages. Despite this, the increase in numerical data was significant.

The tables we produced on the increased use of statistical data in works by geographers show a different movement in physical geography than in human geography. Unlike human geography, from 1939 to 1948 the use of statistical data in physical geography occurs primarily on a regional, state, and municipal scale. Measurements of meteorological stations are very common, especially in a municipal and urban scale. Effectively, the first national syntheses which included measurements, be it on the weather, on the level of vegetation dispersal, or on the shape and altitude of the relief started circulating around 1953 (Figure 10.3). In the journal, the first area to use imagery from measurements is climatology. This was followed by ever more precise relief maps and then by botanical geography. These were the three areas in which there was a greater occurrence of images. Botanical geography appears to be the area that remained connected to the descriptive values of geography the longest.

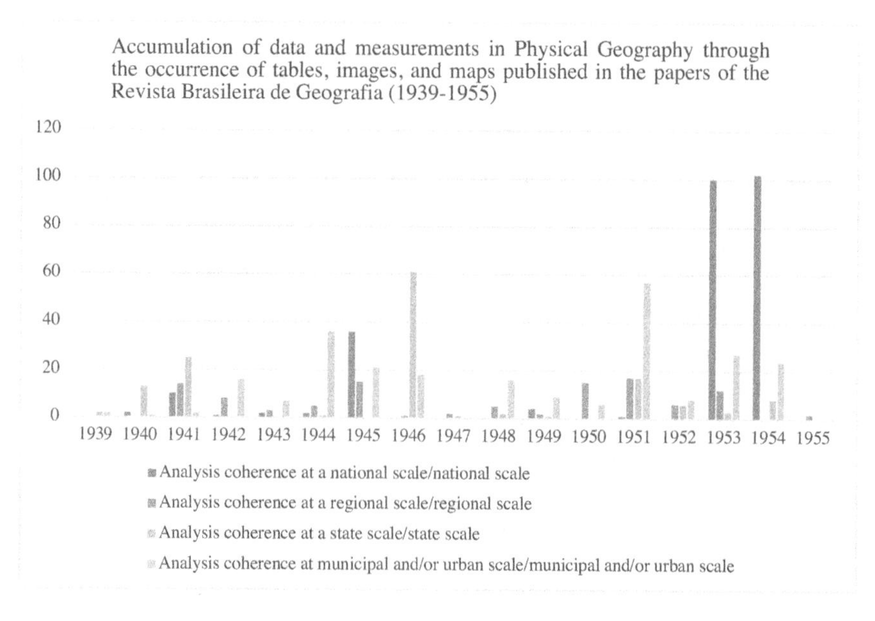

Figure 10.3 Accumulation of data and measurements in Physical Geography through the occurrence of tables, images, and maps published in the Revista Brasileira de Geografia (1939–1955).

Regarding human geography, the first statistical data accumulated, mainly expressed through tables and maps, are national syntheses and analyses. Human geography first elaborates a complete view of the country and some of the first national data is on population density. From 1939 to 1944, the use of statistical data with national coherence prevails over the regional, state, and municipal/urban levels; this is reversed in 1948 (Figure 10.4). As stated, the first syntheses in human geography aimed mainly to calculate population density. After this, statistical data starts to focus on the economy, particularly in terms of the use of the land. Populational movements, in connection with the matter of colonization, follow. It is important to point out that industrial processes are absent in this period (Figure 10.4).

The quality of this information is better represented through images than graphs. I understand image analysis based on one of the lessons of graphic semiology (Bertin 1967), which states that finer, less coarse lines represent more accurate data, while coarser lines represent less precise, geo-referenced data. In general, it can be seen that the IBGE and the CNG advanced quickly in the production of ever more precise maps between 1939 and 1955. A milestone of this process is the publication of a new cartographic basis for Brazil in 1944 followed by a new map of Brazil (Figure 10.5).

Between 1939 and 1942, maps had thicker lines and were usually black and white. From this date on, lines get finer and finer and data is more and more precise from a geo-referencing perspective (Figure 10.6).

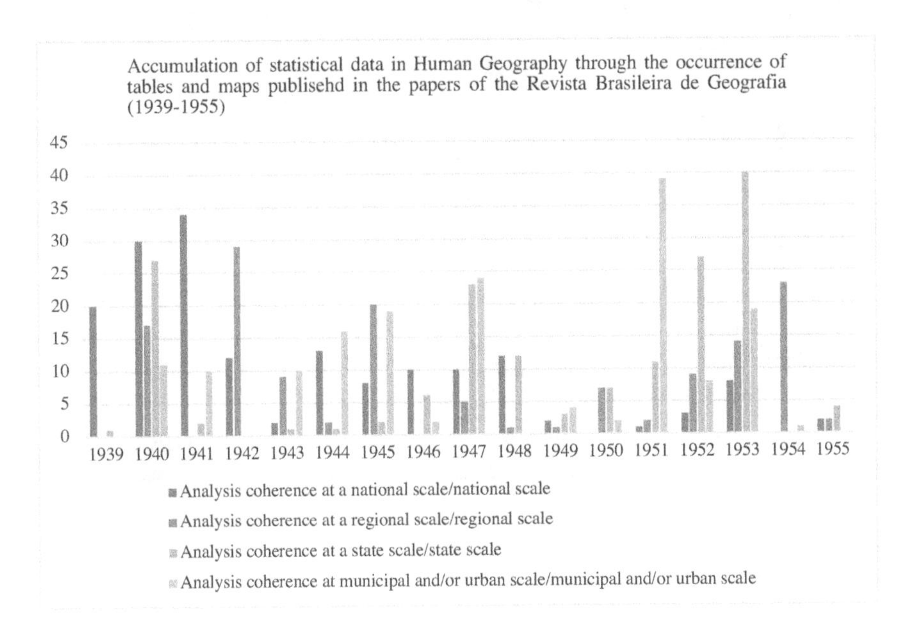

Figure 10.4 Accumulation of data and measurements in Human Geography through the occurrence of tables, images, and maps published in the Revista Brasileira de Geografia (1939–1955).

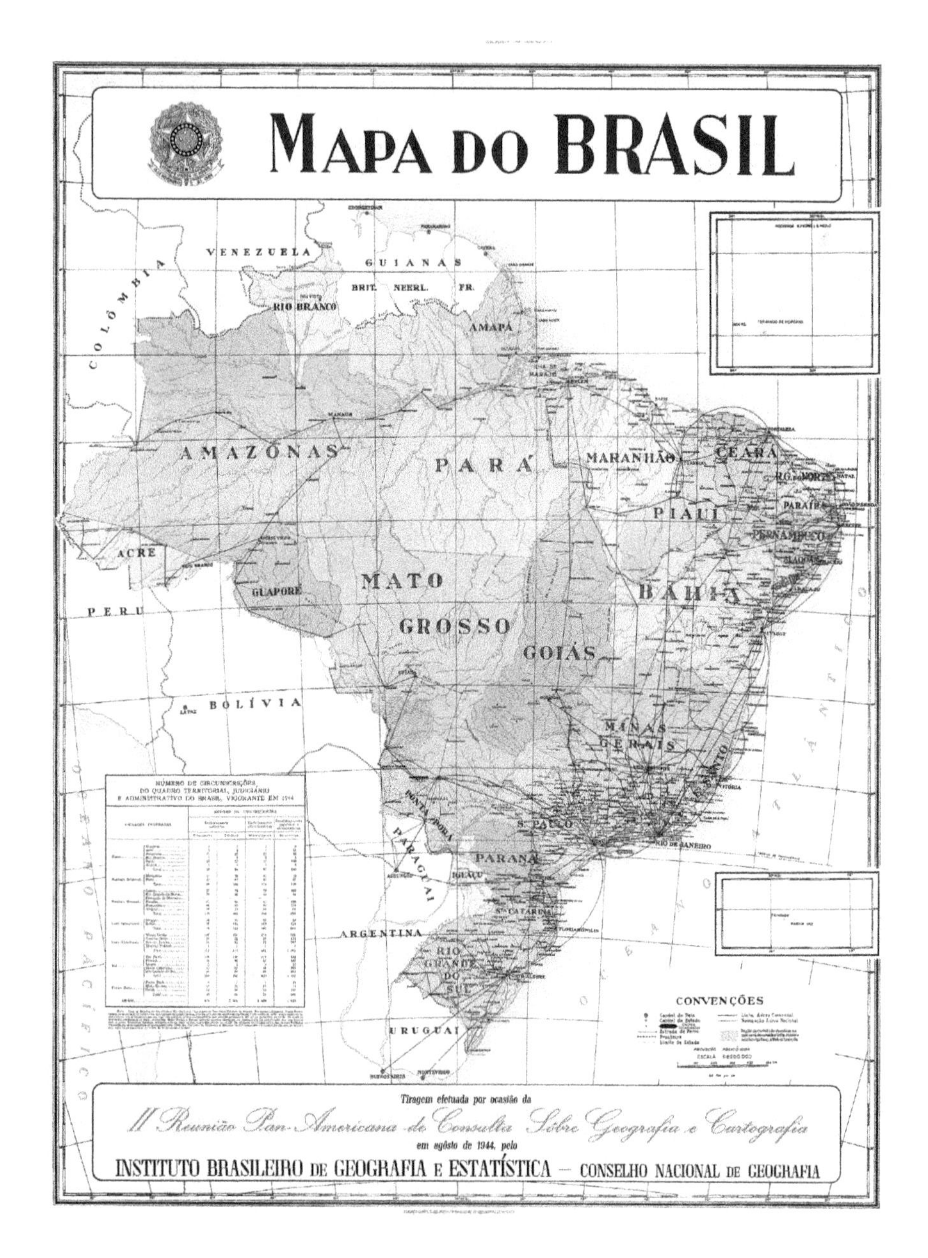

Figure 10.5 Mapa do Brasil (Brazil's [political] map), 1944.

Source: The IBGE digital library catalog

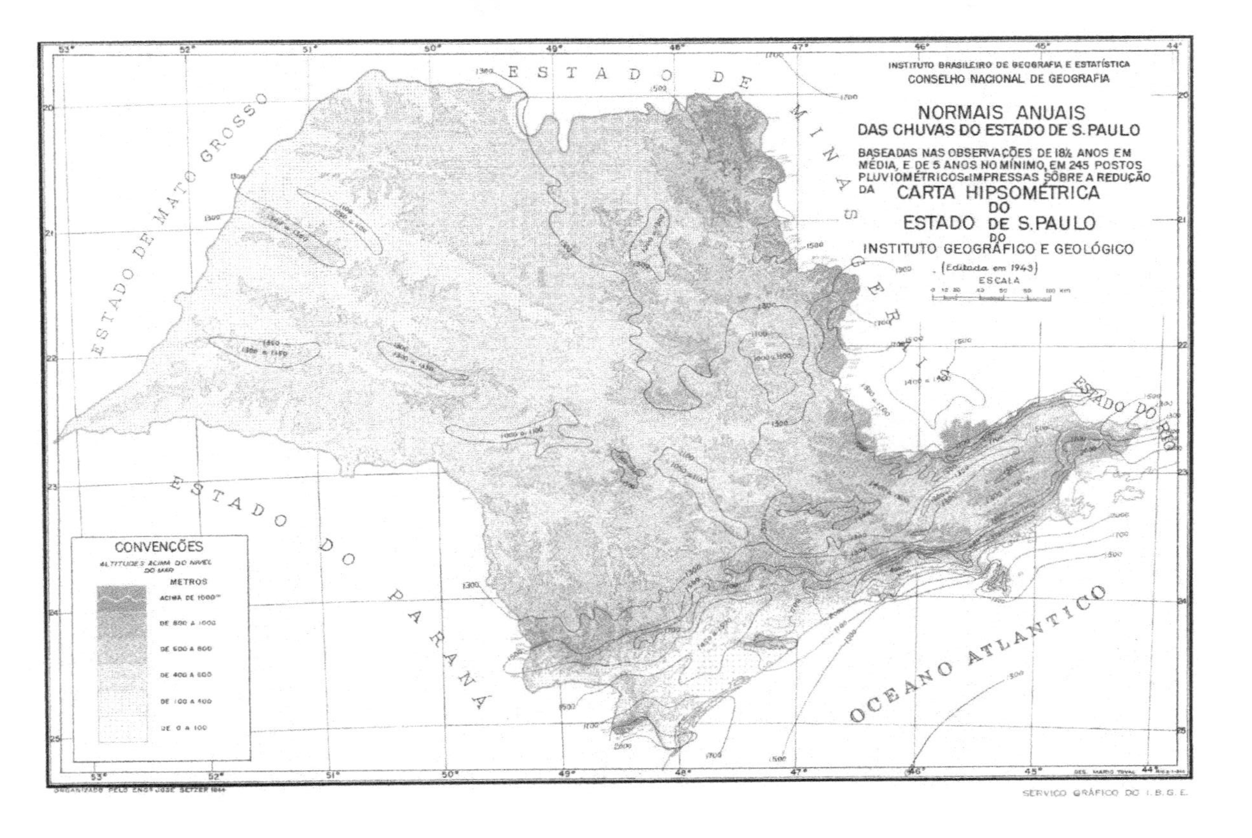

Figure 10.6 “Average curves of annual rain in the state of São Paulo.” This map appears in the journal in 1946. Data on average curves of rainfall are cross-checked with the relief. This information is highly precise.

Source: *Setzer (1946, 64–65)*

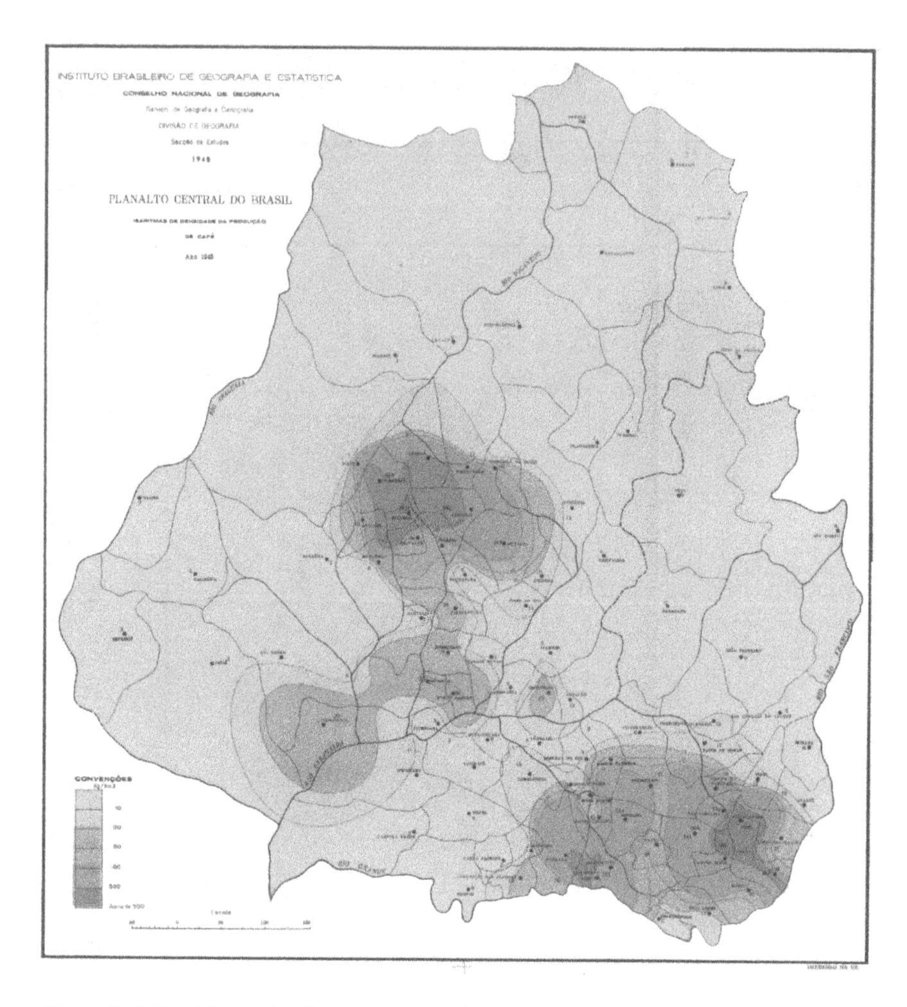

Figure 10.7 "Isarithms of coffee production density". Published in 1950, data representation through isarithms reveals once more the greater precision technicians and geographers from the IBGE and CNG achieved in 1950.

Source: Mello (1950, 80–81)

From 1950 onward, color maps start being made (Figure 10.7). The main criteria for analysis that the IBGE and CNG gathered in this period were: data and maps on weather, relief, vegetation, mineral resources, population density, and colonization movements and land use. In 1959, the first atlas of Brazil is published by the IBGE (Figure 10.8), which we see as the final product of this statistical movement preceding what is understood as the milestone of quantitative geography, 1960. In 1960, greater accuracy can be seen. Data on average curves of rainfall are cross-checked with the relief. This information is highly precise.

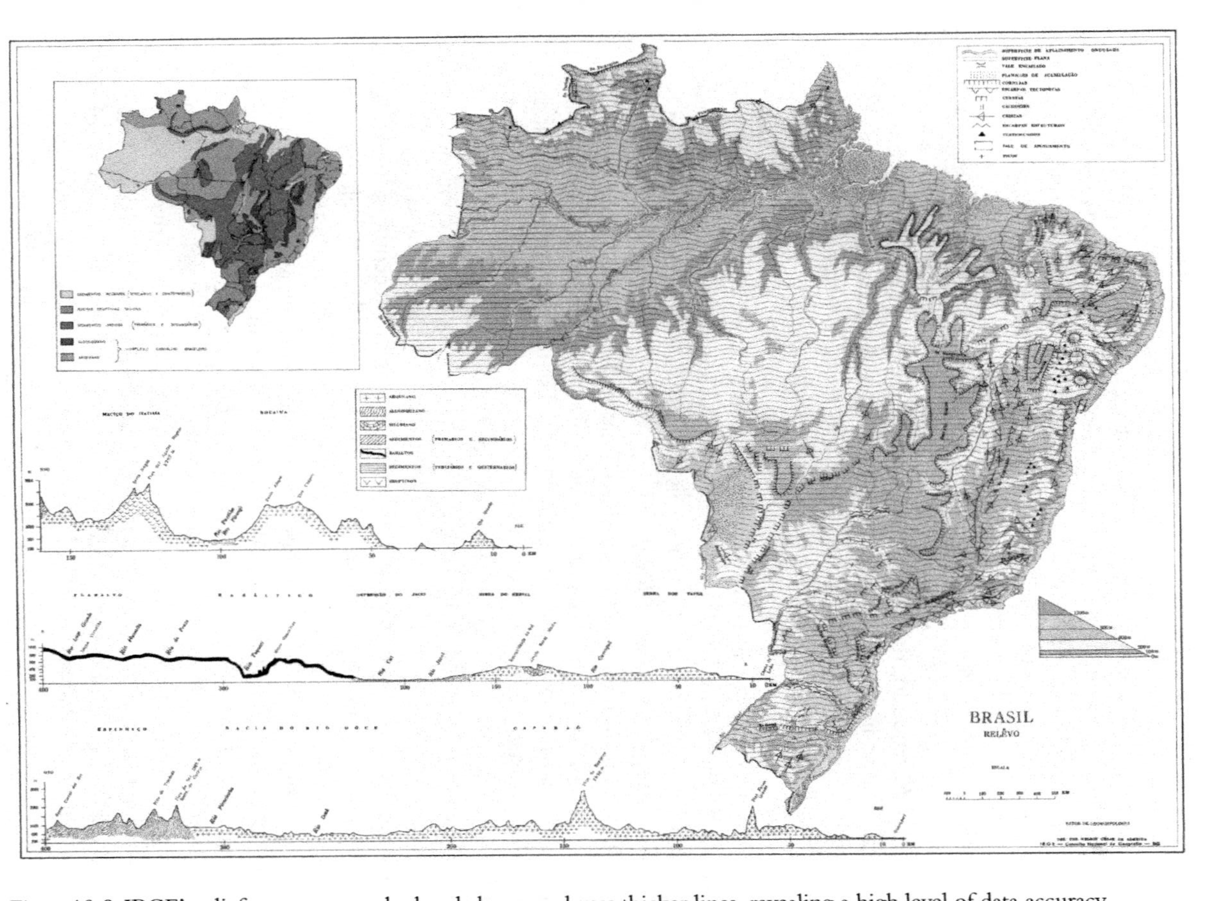

Figure 10.8 IBGE's relief map seems to be hand-drawn and uses thicker lines, revealing a high level of data accuracy.

Source: Conselho Nacional de Geografia (1959)

Countries and plans that did not come to fruition

With these demonstrations, the country started producing statistical data at a considerable pace and rapidly evolving degree of precision and gathered important statistical resources for planning. Between 1939 and 1960, more data were produced on the variables shown earlier. Parallel to these diagnostics, the country created plans based on them. These diagnostics led to a number of conclusions: the country had extensive natural resources; its tropical nature did not prevent it from being populated; the population was excessively concentrated at the coast; its continental nature did not make internal communication easy; and *latifundio* created poverty.

In this manner, the intervention plans that accompanied these diagnostics were: to promote resource valorization through economic planning; promote a vast plan of colonization for the country; review the federative pact and the country's political division in pursuit of more balanced order; and divide property.

At the time of the dissemination of these ideas, geographer Leo Waibel arrived in Brazil, in 1946. Thus, more than just the drive to produce statistics, the moment is marked by how best to use these numbers. One of Leo Waibel's first publications in Brazil and in Portuguese is the article "Von Thünen's theory on the influence of market distance relative to the use of land and its application in Costa Rica" (Waibel 1948). With this article, Waibel traces what should be considered as planning policy for Brazil.

According to Waibel, the theory shows that in an "Isolated State" – which must have fertile land and a population with the same level of education – land price and culture systems vary according to the distance from the market at the center of the state. This being established, the market can regulate prices and function according to a logic of competition. Waibel then analyzes the state of Costa Rica, which would prove Thünen's model even for tropical countries as its main economic tendency would be extraverted. However, as long as some conditions are met, such as the presence of fertile land, an urban center isolated from the agriculture exportation region, and a population of relatively well-educated settlers, markets can regulate the organization of space (Waibel 1948, 32).

In the same year, 1948, Lucio Castro Soares published his article "*Delimitação da Amazônia para fins de planejamento econômico*" (Delimitation of the Amazon for economic planning). His approach focuses on the use of natural resources. In 1947, the *Revista Brasileira de Geografia* published Brazil's general road plan, which aimed to unify the territory from an endogenous perspective on the use of human and natural resources (Silva 1947). In 1949, Hilgard O'Reilly Sternberg published a commentary on Herbert Wilhelmy's work on German colonization of South American forest land, concluding that "of the three groups in which subtropical South-American forest settlers can be divided, the most useful from a farmland conservation perspective is the one that attains the highest degree of soil fixation, meaning the small farmer" (Sternberg 1949, 612).

Between 1948 and 1950, IBGE's geographers also discussed a new plan for mobilizing the population from the shoreline to the countryside through the change of the capital's location, stating that this would lead to an inversion of the importance of latifundia for small properties in the Brazilian economy. Finally, Teixeira de Freitas, IBGE's president at the time, and other geographers and economists proposed and debated a new political organization for Brazil, in which the states would have the same population density, developed internal economy, size, and budget, rebalancing the federative pact. Americo Barbosa followed the general idea of Teixeira de Freitas:

> There was never a preoccupation with dividing the land according to the interests of production, as many non-Iberian countries did in their colonies. This initial mistake has been causing serious woes to the collective, since once all territorial assets of the public power are diluted in private hands . . . making it difficult for a positive policy of economic reorganization to be elaborated.
> [emphasis added] (Oliveira 1947, 35 [translated by the author])

These plans either did not come to fruition or took a great deal more time to be put in place than previously imagined. In fact, all plans seemed to converge on the need to divide the property and were blocked by the political will of the "owners" of the country. Therefore, the numeric diagnostics that Brazil accumulated did not alter its social structure in the short or medium term. Instead, geographic planning based on these statistical diagnostics took second place to the desire of the elites to put a stop to the developmentalist process. When, in 1960, Brazil's new president announced the results of this movement and the necessary implementation of a core reform in Brazil, which included Land Reform, a military coup d'état interrupted the process and in 1964 a dictatorship took hold. In fact, despite the advancements in statistics, the quantitative revolution being planned in Brazil apparently would not lead to social and economic modernization of the territory.

Conclusion

This article attempted to highlight certain facts: the accumulation of numerical data in Brazil occurred late in comparison with Europe but took place relatively quickly by mobilizing modern techniques between 1939 and 1960; however, not only was there a problem concerning statistical production but also this data should be used, thus highlighting epistemological and social issues.

French geographers were more inclined to use these numbers in descriptive, regional, and historical monographs while Brazilian geographers, due to an influence from the German and North American schools of geography, were more inclined to use them for planning. The period analyzed is also that in which Brazilian academic geography is slowly migrating from a strong French influence to a greater Anglo-Saxon and German influence. In addition, statistical diagnoses both from human geography and from physical geography led to the creation of a national perspective on the problems of the country and possible solutions to these problems.

Therefore, in addition to the epistemological perspectives, it was agreed to some extent that the modernization of Brazil might include the division of land properties; at this point, the movement of modernization was brought to a halt by the political desires of an important Brazilian social sector. The social and epistemological consequences of this fact could lead to new areas of research.

Sources

Catalog of the Digital Library of IBGE

Revista Brasileira de Geografia (all the numbers published between 1939 and 1960)

Websites

CLUBE de Engenharia. O Clube de Engenharia na confecção da Carta Geográfica do Brasil para a comemoração do 1° Centenário de Independência do Brasil. Clube de Engenharia Website, 2019. Retrieved from: http://portalclubedeengenharia.org.br/2019/01/11/o-clube-de-engenharia-na-confeccao-da-carta-geografica-do-brasil-para-a-comemoracao-do-1o-centenario-de-independencia-do-brasil/. Accessed: February 27, 2020

Notes

1 Visiting professor at the University of Minas Gerais. Research supported by the Coordination for the Improvement of High Education Personnel (CAPES). I would like to thank CAPES and UFMG for the possibility to carry out this project.

2 Statistics and quantitative revolution are closely linked in this paper for the Brazilian context, even if perhaps this is not always the primary association for the quantitative revolution in general. The dynamics of statistics are currently considered very important for "quantitative geography" as a whole.

3 A positive view of modernization is adopted here. It is important to emphasize that Brazil is a country with a colonial past. Thus, even the so-called progressive Brazilian intellectuals, particularly since 1930, evaluate in complex terms the arrival of capitalist forces to the territory, which bring processes of submission, as well as the possibility to liberate the country from some of the constraints of the colonial system.

4 Pierre Deffontaines (1894–1978) was a French geographer and the disciple of Jean Brunhes, who was an active contributor to geographies in the Americas and part of university missions in Brazil and in Canada as well as Spain. For information on his life and stay in Brazil, see Delfosse 2000.

5 Pierre Monbeig (1908–1987) was a French geographer who participated in the French mission sent to São Paulo in 1935 (see Lira 2021).

6 Francis Ruellan (1894–1975) was a geomorphologist who participated in the French university mission sent to Rio de Janeiro and made important contributions to the IBGE. For more details on this connection, see Aranha 2014.

7 This type of characterization – referring to a French, an American, or a German geography – does not imply a lack of heterogeneity. However, my approach here is closely related to an anthropology of the sciences, evoking long-standing, engrained traditions and cultural areas.

8 The regional scale does not coincide with the administrative scale. In our corpus the scale of analysis corresponded to the biggest and smallest scales of the state dimension.
9 The titles of images and tables were the main references for defining the scale of analysis. For the images, the cartographic grid of reference is a clear indicator for this choice. However, for tables, when a deeper analysis focused on a larger scale (with a smaller unit), we chose to dissociate the data presented in the table.
10 I see the idea of developmentalism as an attempt to promote actions in the country to improve the social and economic conditions of the population as a whole.

References

Adas, S. (2006): *O campo do Geógrafo: Colonização e Agricultura na Obra de Orlando Valverde (1917–1964)*. Doctoral Thesis. São Paulo: Universidade de São Paulo, volume 1.

Almeida, C. M. (1868): *Atlas do Império do Brazil*. Rio de Janeiro: Lithographia do Instituto Philomathico.

Almeida, R. S. (2000): *A Geografia e os Geógrafos do IBGE no Período de 1938–1998*. Doctoral Thesis. Rio de Janeiro: Universidade Federal do Rio de Janeiro, Instituto de Geociências.

Aranha, P. (2014): o IBGE e a Consolidação da Geografia Universitária Brasileira. In *Terra Brasilis* (Nova Série) [Online], 3, put online on 30 June 2014 (accessed 19 February 2020).

Bertin, J. (1967): *Sémiologie Graphique: Les Diagrammes, les Réseaux, les Cartes*. Paris: La Haye, Mouton, Gauthier-Villars.

Braudel, F. (1992): A Longa Duração. In Braudel, F. (Ed.): *Escritos Sobre a História*. São Paulo: Perspectiva, 41–78.

Conselho Nacional de Geografia (1959): *Atlas do Brasil*. Rio de Janeiro: IBGE.

Delfosse, C. (2000): Biographie et Bibliographie de Pierre Deffontaines (1894–1978). In *Cybergeo, Épistemologie, Histoire, Didactique*. Put online on 9 March 2000, modified on 2 May 2007, n.127 (accessed 27 February 2000).

Duarte, R. B. (2013): *Incógnitas Geográficas: Francisco Bhering e as Questões Territoriais Brasileiras no Início do Século XX*. São Paulo: Alameda, FAPESP.

Lira, L. A. (2021): *Pierre Monbeig e Formação da Geografia no Brasil: Uma Geo-história dos Saberes (1925–1956)*. São Paulo: Alameda.

Machado, L. (2012): O. Origens do Pensamento Geográfico no Brasil: Meio Tropical, Espaços Vazios e Idéia de Ordem (1870–1930). In Castro, I. E. de; Gomes, P. C. da C.; Corrêa, R. L. (Eds.): *Geografia: Conceitos e Temas*. Rio de Janeiro: Bertrand Brassil, 309–352.

Machado, M. S. (2000): A Implantação da Geografia Univeritária no Rio de Janeiro. In *Geographia*, 2(3).

Maciel, L. A. (1998): *A Nação por um Fio – Caminhos, Práticas e Imagens da "Comissão Rondon"*. São Paulo: EDUC.

Mello, B. C. C. (1950): Interpretação do Mapa de Produção de Café no Sudeste do Planalto Central do Brasil. In *Revista Brasileira de Geografia*, 12(1).

Miceli, S. (2011): Condicionantes do Desenvolvimento das Ciências Sociais. In Miceli, S. (org): *História das Ciências Sociais no Brasil*. São Paulo: Sumaré, volume 1, 91–133.

Monteiro, C. A. (2002): A Geografia no Brasil ao Longo do Século XX: um Panorama. In *Borrador*, 4.

Oliveira, A. B. (1947): Diretrizes Para Uma Planificação Regional do Brasil: Atividades de Base. In *Boletim Geográfico*, 5(49), abr., 35–41.

Penha, E. A. (1993): *A Criação do IBGE no Contexto de Centralização Política do Estado Novo*. Rio de Janeiro: Fundação Instituto Brasileiro de Geografia e Estatística.

Polanco, X. (1989): Une Science Monde: la Mondialisation de la Science Européene et la Création de Traditions Scientifiques Locales. In Polanco, X. (Ed.): *Naissance et Développement de la Science-monde*. Paris: la Découverte, cap I, 10–52.

Quaini, M. (1983): *A Construção da Geografia Humana*. Rio de Janeiro: Paz e Terra.

Santos, C. J. B.; Castiglione, L. H. G. (2014): A Atuação do IBGE na Evolução da Cartografia Civil no Brasil. In *Terra Brasilis* (Nova Série) [Online], 3, put online on 26 August 2014 (accessed 29 August 2016).

Senra, N. C. (2014): A Junção do G ao E na Formação do IBGE. In *Terra Brasilis* (Nova Série) [Online], 3, put online on 26 August 2014 (accessed 29 August 2016).

Setzer, J. (1946): A Distribuição Normal das Chuvas no Estado de São Paulo. In *Revista Brasileira de Geografia*, 8(1).

Silva, M. M. F. (1947): A Expansão dos Territórios Interiores. In *Revista Brasileira de Geografia*, 9(3).

Sousa, A. N.; Vaz, C. B. (Eds.) (2019): *A Geografia no Alvorecer da República*. Salvador: Editora da Universidade Federal da Bahia.

Sternberg, H. O. R. (1949): A Propósito da Colonização Germânica em Terras de Mata da América do Sul. In *Revista Brasileira de Geografia*, 11(4), 591–612.

Waibel, L. (1948): A Teoria de Von Thünen Sobre a Influência da Distância do Mercado Relativamente a Utilização da Terra e Sua Aplicação na Costa Rica. In *Revista Brasileira de Geografia*, 10(1).

Wyly, E. (2019): *Geography's Quantitative Revolutions: Edward A. Ackerman and the Cold War Origin of Big Data*. Mogantown: West University Press.

11 Italian geographers and the origins of a quantitative revolution

From natural science to applied economic geography

Matteo Proto

Premise: the rise of the quantitative revolution in Italy

In 1970, the geographer Giuseppe Dematteis (b. 1935) published the first systematic paper in Italian addressing the quantitative revolution and the new geography in Italy, starting from its origins and evolution at the international level and continuing by examining its repercussions – more or less overlooked – on Italian geographical sciences (Dematteis 1970).

Dematteis' portrait of Italian geography reveals a classical stereotype of Italian academic thought and Italian contemporary history more broadly: the idea of a permanent failure to innovate, shortcoming, and backwardness as compared to the supranational dimension and international scientific standard. Although quantitative methodologies in the Anglo-American academy had already been critiqued in the early 1970s and were going out of fashion (Barnes 2014; see also Harvey 1972), Dematteis called for a paradigmatic revolution that he argued would push Italian geography into the mainstream of international science:

> Sotto l'aspetto teorico la "rivoluzione quantitativa" si può dunque interpretare come un aspetto di una più ampia "rivoluzione" che investe tutta la geografia. Lo sforzo di creare una nuova tecnologia per la soluzione dei problemi posti dall'organizzazione territoriale ha favorito in forma indiretta nei paesi tecnologicamente più avanzati il distacco del contenuto "scientifico" della vecchia geografia da quello umanistico e il suo organizzarsi nel nuovo indirizzo della geografia teorico-quantitativa.[1]
>
> (Dematteis 1970, 49)

At that time Dematteis was a young scholar at the beginning of his academic career. Just a decade later, he had risen to become a prominent figure in Italian geography by developing a critical reflection on the nature of the discipline, especially in its relationship with power and political practices. In his frequently revisited and most widely quoted (at least in Italy) 1985 book, he also criticized the quantitative methodologies that he had vigorously supported 15 years before, pointing out the theoretical weakness and the poor value of systemic and

DOI: 10.4324/9781003122104-11

functionalist interpretations. In his opinion, such methodological approaches demonstrated the lack of all deductive explanatory power in geography (Dematteis 1985; see also Fall and Minca 2012).

What stands out here, however, is the way in which his 1970 quote underlines the idea of rupture, break, and discontinuity, regarded as implicit in the quantitative revolution he invoked as a remedy for the theorizations (or lack thereof) and methodologies (regarded as unscientific) of the country's consolidated geographical tradition.

In this chapter, I excavate Italian geographical theorization and methodologies from the last decades of the 19th century onward to demonstrate, on one side, that systemic and calculative approaches held a significant place in the discipline long before the quantitative revolution that occurred in Europe and North America after WWII. On the other side, I show the substantial continuity and congruence between the tradition of Italian, natural science-based geography established during the 19th century and the postwar applied economic geography that, as several authors have noted, drove the discipline toward functionalism (Gambi 1964; Farinelli 1980, 2003).

Of course, the rise and consolidation of quantitative analysis in Italian geography followed the spread and dissemination of neo-positivistic approaches occurring in other Western countries and the Soviet Union (Lando 2020). As Trevor Barnes has highlighted (2004), quantitative revolution in Northern Europe and in the United States was led by young scholars who rose up against the academical tradition and particularly, in the American case, against Hartshornian regional geography. On the contrary, in the Italian case, most of the researches acting as pioneers and instigators of the new quantitative path had been trained and taken positions as senior scholars in the interwar period: in fact, the mainstream of geographical research developed without a significant break from the natural positivist perspective established in the second half of the 19th century and substantially unchanged since then, to the functionalist, neo-positivist approach of the post-WWII period. Despite a lack of a deep theoretical thought, as accurately noted by Dematteis (1970), the only switch that actually took place was from methodologies and theoretical frameworks borrowed from the natural and life sciences to those borrowed from mathematics and econometry.

Calculation, statistics, and cartography: the foundations of modern academic geography in Italy

Geographical sciences in Italy emerged as a modern form of knowledge in the 19th century and particularly during the historical revolutionary phase of the so-called *Risorgimento* (ca. 1820–1861), that marks the beginning and the end of the Italian national renaissance and process of political unification. As stressed by historians (Pecout 1997; Casalena 2007) and historical geographers (Gambi 1973; Sturani 1998; Proto 2014a), geographical images and discourses capturing the conceptualization of the country were exploited as a neutral, quantitative instrument for stressing the very idea of a unified nation while national unity itself was generated by privileging distinctive demographic and sociocultural features over precise, circumscribed territorial boundaries. This occurred first of all through the enumeration and calculation of various countable data related to both physical and human geography. The main goal of import for the so-called *Statistici* (statistician scientists) – mainly geographers, such as Adriano Balbi (1782–1848) and Cesare Correnti (1815–1888) among the others – was to depict an image of the country in a revolutionary frame. The purpose of such a framing was, on the one side, to circumvent political censorship through the depoliticization of their discourse and, on the other side, to imbue apparently impartial scientific and quantitative observations with an implicit political significance. This same strategy was also embedded in the establishment of modern German geography by Alexander von Humboldt at the beginning of the century (Farinelli 2000), founded on the bourgeois pretense of separating the moral dimension from the exercise of power (Koselleck 1959).

These methodologies made a significant mark on the evolution of Italian geography, especially in the decades from 1870 to 1900, when geographical knowledge was established and consolidated in the academy following Italian national unification. Meanwhile, since mid-19th century a strong positivistic approach had been conditioning scientific thought, fueled by Charles Darwin's evolutionism and more broadly through the adoption of models and methodologies reshaped by the natural and life sciences, powerfully dominated by statistics and cartography, and applied to the humanities and social sciences. This tendency, common in geography at the international level as well (Rupke 2011; Jureit 2018) was particularly weighty and long-lasting in Italy.

One example of this theoretical and methodological approach can be seen in the understanding and definition of geographical region systematized at the turn of the century by Giovanni (1846–1900) and Olinto Marinelli (1876–1926), two leading players in the establishment of modern academic geography in Italy: an idea of region strongly based on a topographical surveying and representation of both physical-natural features (morphology, climate, flora, fauna, etc.) and anthropic ones (demography, settlement, etc.). In this context, the enumeration and quantification of data played a fundamental role in depicting the distribution of geographical elements in space, which in turn made it possible to engage in multifaceted aggregation in order to identify the region in all its complexity, as the result of interaction between different phenomena and as a synthesis of the humanity–nature relationship (Proto 2014b; see also Marinelli 1916a). This apparently neutral and pure, scientifically generated model became a powerful instrument for defining geographical borders, a definition which was then exploited to assert a scientific understanding of Italian national borders and political space on the eve of the WWI (Proto 2017). Moreover, despite the fact that its investigation was mainly based on natural and life sciences methodologies and not of mathematics, such a regional model – defined *regione integrale* (integral region) – was telling similar to the systemic, functional region as defined in quantitative geography. The similarity stemmed from the fact that, according to positivist Italian geographers, theoretical models and representations were not merely a simplified and incomplete substitution for direct observation, but the only way to contemplate and consider the vastity and the complexity of the terrestrial surface with its socio-natural interactions (Marinelli 1902).

Along these lines, it is important for my purpose here to stress that cartography played a key role as both a representational tool and an analytic source for geographical surveying, resulting in the intense systematization and modelling of geographical knowledge. This process peaked with the conceptualization of *tipo geografico* (geographical type) formulated by Olinto Marinelli: geographical types were abstract formal models, inferred from topographical maps, that allowed researches to conduct comparative studies of both physical-natural and anthropic topographical traces and, as such, tools of a form of *a posteriori* knowledge (Proto 2014b; see also Marinelli 1916b). This approach allowed Marinelli to develop a general theory so as to infer – by experimenting with its circumstantial application – the very truth of the theory itself. Geographical elements and phenomena therefore had to be situated within a model that predated the act of identifying them "con più di mezzo secolo di anticipo sulla geografia quantitativa[2]" (Farinelli 2003, 128).

It is not a coincidence that Olinto Marinelli also played the role of launching quantitative urban studies in Italian geography. In 1916, he published an article in the field of urban geography that presented the results of a comparative study of American and Italian cities, conducted following his involvement in the Transcontinental Excursion promoted by the American Geographical Society under the mentorship of William Morris Davis. This paper had the significant title of *Dei tipi economici dei centri abitati. A proposito di alcune città italiane e americane* (On the economic types of settlement: Investigating some Italian and American cities). By reflecting on the economic role of the cities and their topographic location, Marinelli developed the concept of *città completa* (complete city) to identify the settlements that are situated at the top of hierarchical positioning in space (Marinelli 1916c). Later on, the German geographer Walter Christaller recognized that this conceptualization of complete city corresponded to his *zentrale Orte* (central places), quoting Marinelli's 1916 work (Christaller 1968 [1933], 23).

Marinelli's regional explanations as well as his topographical ideal modelling became a significant theoretical frame for the further development of Italian geographical thought. Central was the assumption that from an abstract form – like the geographical type inferred from topography – or from a theoretical model – like an idea of region as the result of the interaction of calculable and representable elements – it was possible to infer an explanation about the nature of the processes and the functioning of the world.

The interwar period: empowering of the state through the implementation of applied sciences

Positivist and natural science-oriented geographical theorizations were then exploited after WWI, when the discipline of academic geography underwent a process of professionalization and its practitioners produced series of applied research works, powerfully supported by the state. In this context, the *Consiglio Nazionale delle Ricerche* (National Research Council), a newly established research hub created by the government to manage scientific research outside of the universities, played a significant role. Although this organization was not founded until 1923, its origins date back to the WWI period as an effort to coordinate and promote science in order to improve industrial development and national security. After the rise of the Fascist dictatorship, this public body became mainly a powerful instrument in the hands of the regime that was used to pursue its political and governmental goals. The Comitato Geografico Nazionale (National Geographical Committee), founded in 1921 to coordinate geographic research in the country and connect Italian geography with the transnational network of the newly founded IGU, also merged into the National Research Council (Gambi 1994; Martelli 2001).

With the consolidation of the regime, geographical surveying was also increasingly conditioned by governmental interests and driven especially toward a utilitarian understanding of research. This development entailed the specialization of research fields and a tendency toward applied approaches. More broadly, geography lost the universal and generalist dimension it had during late 19th century to become more similar to applied studies.

On the many geographical research projects sponsored by the National Research Council between the end of the 1920s and WWII, three topics are particularly illustrative of this new trend as well as the methodological shift toward quantitative and analytic frameworks. Two of these had to do with the investigation of human settlement and involved questions of agrarian geography and nonurban areas. The first survey consisted in monitoring the mountain depopulation processes that had been taking place on a massive scale since the end of the 19th century. This topic was particularly significant for the regime and specifically for its anti-urban, pro-agrarian rhetoric. Given the obvious political risks, however, researchers were obliged to avoid any sociohistorical investigation or references which would have delved into the reasons – including political ones – behind this phenomenon of migration away from mountain areas. Studies thus concentrated on demographic dynamics and economic issues by enumerating statistical data in an effort to formulate quantitative models to explain the tendency and evolution of the phenomenon (Giusti 1938).

The second topic, the classification of Italian rural houses, was imagined as an ethno-cultural research project investigating the functional structure of rural settlement. By classifying and analyzing the different typologies of rural houses according to a structural framework and isolating their functional components, this survey sought to define a general model of Italian rural housing so as to characterize and preserve the essence of the rural Italian landscape. The types were mainly based on topographical and formal elements that tended toward abstraction, as in Marinelli's geographical type, and avoided any ethno-historical considerations. These isolated functional elements were then also investigated in terms of diffusion and circulation to identify certain regions in which certain formal models were widespread. The research project soon merged with a number of nationalist and chauvinist discourses, as the appearance of typical Italian elements in rural houses – e.g., in the newly conquered lands, after WWI – was understood as an expression of Italian civilization. This position was sustained, for example, by Bruno Nice (1916–1993) in his investigation of rural houses in the Julian Venetia, namely the Trieste region and the Istria peninsula, an area characterized by a mixed population of Italians, Slovenians, and Croatians. The research project was then connected to a grand national project of recovering and rebuilding rural houses and to resettling the rural population, a project which in the eastern regions came to focus more and more on the Italianization of non-Italians. As Nice himself clearly stated:

> Perciò l'opera di redenzione delle case rurali nella Venezia Giulia deve avere un fine di più rispetto agli altri compartimenti del Regno: la conservazione e la restaurazione della fisionomia italiana degli insediamenti e delle abitazioni rurali nella terra che costituisce una nostra testa di ponte nell'Europa centrale e nella Balcania.[3]
>
> (Nice 1940, 137)

The term redemption – which Nice employs to characterize the preservation of Italian rural houses – is a meaningful keyword that appears quite frequently in this period, including in political speeches and propaganda. This political meaning of *redenzione* replaced the word's theological meaning to become a synonym for the conquest, purification, and restoration of land.

First, it involves a distinction between unredeemed and redeemed lands, with the first indicating Italian regions under a foreign state and the second the areas that has been returned to fatherland but required intervention to restore their original cultural-national character. Redemption was also very commonly associated with wetland drainage and land improvement, however, as in effort to combat malaria. In the interwar period, therefore, redemption came to indicate the struggle to strengthen the nation, its economy, resources, and features. Italian geographers' contribution to this nationalist goal was similar to the role German geographers played, in the same period, in the newly conquered eastern lands inhabited by Slavic peoples (Barnes and Minca 2013).

The last highly significant research topic comprised a host of surveys dedicated to port management and maritime transportation, which were specifically focused on the intra- and extra- European traffics in Italian ports. This issue was very important for the regime as it pushed to relaunch maritime politics in terms of bolstering both naval military power and the country's trade and economic power (Frascani 2008). Geographical researchers focused especially on the ports of Naples and Genoa with the aim of developing them and transforming them into European and colonial hubs. In these surveys, mainly based on statistic-economic data, geographers tried development model analysis to determine and quantify each port's area of influence, both inland and through its connective network of destinations (Jaja 1936).

The key point here is that, in the interwar period, geographical research merged with the demands of the Fascist regime, especially around issues of human settlement and economic geography. The approach remained very superficial and was confined to topographical descriptions, economic and demographic quantifications and attempts to define certain functional models. Such an approach served, on one hand, to conceal any references to the sociopolitical context and, on the other, to reinforce the state and the regime's politics of power.

Economic geography and the functional region after WWII

After the fall of the Fascist regime in June 1943, a process of purging began that also affected the academic positions and scholars who had been involved in the regime. This highly contested process proceeded in the following years but was then largely discontinued soon after the end of the war, due in part to the large-scale compromise of the ruling class and bureaucratic and administrative bodies of the state – including the educational and scientific organizations – forged with the dictatorship (Domenico 1991). Beyond some temporary suspension from teaching and academic activities, most scholars, even those whose work was compromised by racist and ultranationalist theorizations, were able to maintain their positions by adapting to the new political environment. In the field of geography, it is very significant that some university professors who had already held key positions before the war and showed a particular interest in political geographical issues moved rapidly toward more neutral statistical-economic surveys, pushing for the improvement of quantitative methodologies and approaches. A similar situation can be seen in post-Nazi Germany, where several geographers involved in the regime moved to applied geography (Wardenga et al. 2011).

Of these academics, a significant figure was the aforementioned Bruno Nice. Soon after the war, Nice recovered and consolidated his interest in the applied geography of settlement studies (Nice 1946, 1947), now focused on urban and regional planning (Nice 1952). In 1953, he published a significant book on the relationship between geography and planning in which – probably for the first time in Italian – he defined geography as a spatial science. In his conceptualization and distancing himself from his teacher Renato Biasutti (1878–1965), geographical investigation synthesis was no longer related to the landscape but to space itself:

> considerato da un punto di vista concreto e funzionale, ossia per la vita economica e sociale che vi si svolge con una serie di fenomeni, quali, ad esempio, lo sfruttamento delle risorse naturali e la circolazione dei beni e delle persone.[4]
>
> (Nice 1953, 11)

His theorization began with a critique of French human geography and in particular of Vidal del La Blache and Demangeon's concept of *genre de vie* on the ground that it was superficial and incapable of explaining the complexity of socioeconomical phenomena. In fact, French human geography had not registered any significant impact on Italian geography at that time, and regional surveying was still based on Marinelli's natural-scientific and cartographic approach. Nice also saw an opportunity to update the still-prevailing regional conception that was mainly based on natural sciences with a new regional understanding based on economical facts, particularly the modelling of mobility and communication. In his view, such an approach would make it possible to identify functional regions, distinct from traditional "formal" regions and formed by the circulation and effects of socioeconomical networks.

In this book, Nice cited the work of Von Thünen and Christaller and called for the establishment of stronger interdisciplinary relations between geography and planning. In his mind, the common grounds between these arenas lay in cartography and, more broadly, the idea that planning was in fact the applied version of geography, that is, the materialization of its abstract theoretical models.

The first geographer to introduce the international debate on quantitative modelling and deductive research methodologies in Italy was Eliseo Bonetti (1910–2005), a highly prolific scholar whose work in the afterwar period began to show a marked interest in the localization of economic activities, particularly with regard to the service industry, and developing conceptualizations such as centrality and urban hierarchy. Bonetti spent his entire career at the University of Trieste where he became professor of economic geography in 1965. One of his first tasks was methodically introducing Walter Christaller's theorization (Bonetti 1964) into the Italian academic word, as Christaller's work had not yet been translated into Italian. Starting from the discussion and assimilation of quantitative models, in the following years Bonetti focused on retail trade localization (Bonetti 1967), the hierarchical distribution of settlement (Bonetti 1969), and spatial analysis modelling and theorization (Bonetti 1976).

His geographical education and research activities in the early years of his career reveal a significant and distinctive background. Bonetti had solid ties to the Trieste school of geopolitics that arose in the late 1930s in the wake of Karl Haushofer's German geopolitics and *Geopolitische Zeitschrift*; indeed, the Italian journal *Geopolitica* represented a kind of original derivation of Haushofer's periodical (Antonsich 2009). One of the key, guiding figures of this political and scientific field was Giorgio Roletto (1885–1967), Bonetti's teacher and mentor. Roletto perceptibly conditioned Bonetti's education and initial research interests: imperial political geography and geopolitical topics such as the condition of the Italian colonial empire (Contento and Bonetti 1936), the geopolitics of the east space and Eurasia, and multiple other topics related to space and politics in different world regions.

At the same time, Bonetti began to cultivate a strong interest in economic and social geography (Bonetti 1942a) with a specific focus on policies of port management and trade (Bonetti 1942b), becoming skilled in the collection and processing of statistical data and quantitative analysis. As mentioned earlier, port management had represented a significant geographical research topic since the late 1920s.

In the postwar period, Bonetti continued to pursue his interests in political geography and geopolitics, especially in relation to the Italian northeastern border after the war (Bonetti 1947), before rapidly abandoning such lines of inquiry to take up more neutral topics in the framework of economic geography.

A similar concern with port management and trade surveying also gave rise to the research of Umberto Toschi (1897–1966). Toschi graduated in Bologna in 1921 and became professor of economic geography in Bari in 1935 and subsequently dean of the faculty of economics. In 1951, he was appointed director of the geographical institute in Bologna, a position he maintained until his death. In the late 1950s, Toschi developed a major research program based on abstract system modelling and quantitative methodologies, much more concrete and applied than Bonetti's mainly theoretical forays into this field. Toschi had a somewhat dark past as well, stemming from work he conducted in political geography and geopolitics in the interwar period and then largely abandoned. From the 1930s onward, he developed theoretical studies in the field of political geography from a strongly nationalistic perspective, exalting the figure of Mussolini and his role in establishing the Italian empire overseas. He also introduced an explicit racist perspective, promoting the idea of racial divide in opposition to ethnic mixing and thereby supporting the establishment of the racial laws in Italy (Toschi 1937, 1939). Meanwhile, he developed an interest in urban and settlement studies and the localization of industrial facilities under an economic and quantitative perspective, a development that also entailed introducing and translating into Italian Alfred Weber's work *Über den Standort der Industrie*.

What is the connection between these different topics? Toschi, like many of his contemporaries in academia, subscribed to the organicist understanding of the state developed by Friedrich Ratzel. He clearly explains this perspective in his understanding of the political geography of the state: in living organisms, blood is circulated by means of a circulatory system, a complex, orderly, unitary, and functional system. Adopting an organicist perspective, this systematic framework could be applied to different circulatory systems such as rail and road networks, shipping, postal services, and telephone and radio communication, that is, all the systems on which the functioning of the modern state depended (Toschi 1937). In fact, the question of the nature of regional constitution and influence in shaping the production of goods and their circulation lies at the origins of economic geographical thought (Barnes 2001).

Beginning as early as the interwar period and mainly under the influence of German geographers, economists, and sociologists, therefore, Toschi began to investigate the networks and circulatory systems that shape the geographical space while also calling into question explanatory models throw quantitative research.

After the war, political discourse was neutralized and marginalized in favor of stressing economic and quantitative dimensions and drawing closer to urban studies. This latter became Toschi's main research topic. In 1954, together with Francesco Brambilla, a professor of economics specialized in statistics and econometry, he edited a volume that can be regarded as one of the first examples of quantitative geographical surveying in Italy. The book presents the results of a survey analyzing the economic role of the city of Ivrea and aimed at determining its region of influence. Ivrea was the seat of the Olivetti factory, Italy's largest and oldest typewriter producer and, at that time, also a leader in the production of calculators and teleprinters, what would later become the Italian informatics industry.

The research was aimed at identifying Ivrea's zone of influence by defining the vectors of flow circulating outward from the city. In order to establish these trajectories, they considered demographic data such as migrations and marriages in the area under analysis as well as multiple socioeconomic data. Such data included the provisions of various commercial activities (food stores, pharmacies, local markets) and people consumption of cultural activities (cinema and other forms of entertainment). Through this survey they were able to specify the city's region of influence, a functional region that differed significantly from both the historical region and administrative districts (Toschi and Brambilla 1954).

This work thus constituted the foundation of a new relationship between Italian geographers and the public sphere, a new kind of approach that from the early 1950s onward characterized Italian geographical surveying in supporting the interests of the ruling class and establishment. In the postwar period Italian economy, the state took a leading role in seeking to address and coordinate both the industrial production and consumption through economic planning and interventionism (Salvati 1982). This role can be seen, for example, in the establishment of several ministerial committees, such as for urban planning (from the beginning of the 1950s) and regional development (from the following decade). Geographers' roles in these committees reflected a technical and top-down approach limited to implementing the will of the government by purging socioeconomical processes of their political implications and pushing for a better efficiency in capitalist terms (Gambi 1973). In this context, therefore, the role played by Italian geographers deserves the same critiques that in the late 1960s were addressed to the rise of a critical spatial theory in opposition to technical socio-spatial construction, aimed at supporting capitalist growth and the interests of the ruling class (Castells 1972; Lefebvre 1974).

Epilogue: a critique of the quantitative revolution in Italy

As I have tried to demonstrate, since the beginning of the 1950s certain innovative theoretical and methodological approaches arose in Italian geography that were not too dissimilar from the coeval dominant paradigm, mainly connected to applied economic geography and aimed at empowering the quantitative and modeling approach that had emerged previously with the rise of positivism at the end of the 19th century.

My overview begins from Eliseo Bonetti, the first to introduce Christaller's theorizations in regional studies, and follow the progressive acceptance of functional region and territorial system conceptualizations, based on the researches into polarization processes developed by Umberto Toschi (Vallega 1980).

The actual extent to which Italian geographers used quantitative modeling was rather limited throughout the 1960s as a proportion of geographical surveying as a whole; surveys continued to employ mainly traditional methodologies, in part because Italian geographers lacked the necessary competences in mathematics and did not have access to more sophisticated calculating instruments (Turco 1980). Some significant exceptions are Giuseppe Dematteis and his application of Christaller's grid, as reinterpreted by Brian Barry, to analyze the urban network of settlement around the city of Torino (Dematteis 1966). Later, during the 1970s, empirical case studies of quantitative models and methods enjoyed broader circulation and were utilized more frequently, for instance in spatial analyses of workers' mobilities (Vlora 1977) or the demographic dynamics related to urban centers (Bottai et al. 1978). Moreover, there were some significant theoretical discussions about employing and formulating models and indicators (Dematteis and Vagaggini 1976; Zanetto 1979).

The dominant tradition of Italian geography clashed with emerging perspective during the 21st Italian Geographical Congress in 1971 where, in the context of a crisis of geographical knowledge and the academic tradition of geography, participants engaged in a debate about the paradigmatic character of the discipline. Since the end of the 1960s, Italian geography and academic scientific knowledge more broadly had been undergoing a period of profound turbulence and contestation, due on the one side to the 1968–69 student protests and, on the other side, to the university reforms the government was passing in effort to respond to the demands posed by the students and civil society as a whole (Turco 1980). Moreover, the mainstream of academic geography had fallen under increasing paradigmatic attack since the end of the 1950s, fueled by the antidogmatic figure of pioneering humanist geographer Lucio Gambi (1920–2006). Indeed, on several occasions, Gambi had criticized the still-dominant positivist paradigm from an original historicist perspective forged of multiple elements, including the French school of *Les annales* and the Italian neo-idealist tradition of Benedetto Croce's historicism and Antonio Gramsci's historical materialism (Sofia 2020).

In this context, perspectives derived from Anglo-American discussions and introduced by Dematteis through his provocative 1970 publication cited at the beginning of this chapter entered into the Italian debate about quantitative geography's paradigmatic turn. As stated earlier, quantitative geography had close links to the traditional positivistic paradigm of Italian geography; at the same time, it dictated a theoretical reflection on the nature of geography and its epistemology, a reflection which was wholly missing from Italian work due, as Dematteis argues in his essay, to the restrictive dominant paradigm (Dematteis 1970).

At the 21st congress, the debate over theoretical and methodological issues was launched by Giorgio Valussi (1930–1999). Valussi argued that it was time to adopt quantitative geographical principles in the mainstream of geographical research, thereby resolving the seemingly deep divide between positivist and neo-positivist theories and methods. It was assumed that quantitative geography could be assimilated into the dominant paradigm of natural-scientific geography. This point fact sparked the explosion of the crisis of the traditional paradigm itself, however (Turco 1980). Indeed, Dematteis' challenging arguments had triggered a reflection on the existence of the models embedded in geographical thought, suggesting that geography needed to interrogate the nature of its schematizations. This point gave rise to the humanistic critique of the old paradigm as an analysis of its underlying structure, what lies behind its prevailing theoretical models. And, above all, the need to investigate the relationship between knowledge and power and the ideological substratum intrinsic to geographical research. At the same time that quantitative geography was integrating into the traditional mainstream, therefore, the emerging critical discourse rejected quantitative models on the grounds that "soltanto a patto di rinunciare al paradigma neo-positivistico di modello geografico è possibili comprendere il reale rapporto tra ideologia e paradigma stesso, cioé, in altri termini, tra politica e scienza[5]" (Farinelli 1980, 1025).

In conclusion, what took place throughout the 1970s was a progressive critique of the role geographical knowledge played in sustaining and neutralizing political discourse. Critics stressed that the discipline had not effectively detached itself from 19th-century nationalist bourgeois geography (Dematteis 1985) while interrogating geography's role under the Fascist regime and post-war capitalist technocracy, as Gambi had done since the 1960s (Gambi 1964). It is telling that Gambi launched his newly established human geography series issued by the Milanese publisher Franco Angeli at the beginning of the 1970s with the following preface, still relevant today:

> A idee più chiare non ha portato neanche la cosiddetta geografia quantitativa . . . che pure enunziandosi come nuova e dinamica direzione tesa ad innestare un tipo di ricerca tradizionale nella metodologia e nei problemi odierni, non può nascondere però la sua intima natura: che è quella dei processi che mirano a metamorfosare (neutralizzandoli come valori storici) gli uomini in automi e le realtà umane in artifici, in funzione dei piani operativi d'una ristretta classe di tecnocrati.[6]
>
> (Gambi 1971, 3)

Acknowledgment

I want to thank Franco Farinelli for his suggestions and comments, and especially for illuminating me about certain issues at the origins of critical thinking in Italian geography during a long-distance conversation we had during the COVID-19 pandemic lockdown. I wish also to thank Angelina Zontine for proofreading the chapter.

Notes

1 From a theoretical perspective, the "quantitative revolution" should be understood as part of a broader revolution that is affecting geography as a whole. The need to forge a new technology for the resolution of issues caused by territorial planning in the most developed countries has indirectly supported the move to separate the "scientific" content of old-style geography from the humanistic one, and the rise of a new theoretical-quantitative path in geography.
2 More than half a century before quantitative geography.
3 For this reason, the redemption of rural houses in Julian Venetia has a further purpose as compared to the other regions of the kingdom: the conservation of Italian features and the restoration of the villages and rural houses in this land that represents a bridgehead in Central Europe and the Balkans.
4 Considered from a tangible and functional point of view, i.e., in terms of the social and economic life that takes place in the space and that is related to a range of phenomena such as natural resources exploitation and the circulation of goods and people.
5 Only by giving up the neo-positivist paradigm of geographical model would it be possible to understand the effective relationship between ideology and the paradigm itself or, in other words, between politics and science.
6 Not even so-called quantitative geography led to better ideas; despite its understanding as a new dynamic aimed at introducing a traditional research approach for contemporary problems, it could not hide its intimate nature, i.e., processes that aim to transform human beings (destroying their historical value) into androids and human works into artifacts, in keeping with the operating plans of a delimited technocratic class.

References

Antonsich, M. (2009): Geopolitica: The "Geographical and Imperial Consciousness" of Fascist Italy. In *Geopolitics*, 14, 256–277.
Barnes, T. (2001): "In the Beginning Was Economic Geography": A Science Studies Approach to Disciplinary History. In *Progress in Human Geography*, 25, 521–544.
Barnes, T. (2004): Placing Ideas: Genius Loci, Heterotopia and Geography's Quantitative Revolution. In *Progress in Human Geography*, 28, 565–595.
Barnes, T. (2014): What's Old Is New, and New Is Old: History and Geography's Quantitative Revolutions. In *Dialogues in Human Geography*, 4, 50–53.
Barnes, T.; Minca, C. (2013): Nazi Spatial Theory: The Dark Geographies of Carl Schmitt and Walter Christaller. In *Annals of the Association of American Geographers*, 103, 669–687.
Bonetti, E. (1942a): I Postulati Della Geografia Sociale. In *Geopolitica*, 6, 2–12.
Bonetti, E. (1942b): Il Porto di Fiume. In *Geopolitica*, 3, 4–10.
Bonetti, E. (1947): Il Confine Italo-Jugoslavo Secondo un "Neutrale". In *Rivista Geografica Italiana*, 54, 42–46.
Bonetti, E. (1964): *La Teoria Delle Località Centrali*. Trieste: Arti Grafiche Smolars.
Bonetti, E. (1967): *La Localizzazione Della Attività al Dettaglio*. Milano: Giuffre.
Bonetti, E. (1969): *Centralità e Gerarchia Delle Località*. Milano: Giuffre.
Bonetti, E. (1976): L' Analisi Fattoriale e le Sue Applicazioni Nella Ricerca Geografica. In *Cultura e Scuola*, 60, 152–163.
Bottai, M.; et al. (1978): Analisi Tipologica del Comportamento Demografico dei Comuni Italiani. In *Rivista Geografica Italiana*, 85, 321–347.
Casalena, M. P. (2007): *Per lo Stato per la Nazione: i Congressi Degli Scienziati in Francia e in Italia, 1830–1914*. Roma: Carocci.
Castells, M. (1972): *La Question Urbaine*. Paris: Maspero.

Christaller, W. (1968 [1933]): *Die Zentralen Orte in Süddeutschland*. Darmstadt: Wissenschaftliche Buchgesellschaft.

Contento, A.; Bonetti, E. (1936): *L'impero Coloniale Italiano nei Suoi Aspetti Geopolitici*. Trieste: Istituto Coloniale Fascista.

Dematteis, G. (1966): *Le Località Centrali Nella Geografia Urbana di Torino*. Torino: P. Conti.

Dematteis, G. (1970): *Rivoluzione Quantitativa e Nuova Geografia*. Torino: Arti grafiche Rosada.

Dematteis, G. (1985): *Le Metafore Della Terra: la Geografia Umana tra Mito e Scienza*. Milano: Feltrinelli.

Dematteis, G.; Vagaggini, V. (1976): *I Metodi Analitici Della Geografia*. Firenze: la Nuova Italia.

Domenico, R. P. (1991): *Italian Fascists on Trial, 1943–1948*. London: The University of North Carolina Press.

Fall, J.; Minca, C. (2012): Not a Geography of What Doesn't Exist, But a Counter-Geography of What Does: Rereading Giuseppe Dematteis' Le Metafore della Terra. In *Progress in Human Geography*, 37, 542–563.

Farinelli, F. (1980): Dibattito. In Corna-Pellegrini, G.; Brusa, C. (Eds.): *La Ricerca Geografica in Italia 1960–1980*. Varese: Ask, 1024–1026.

Farinelli, F. (2000): Friedrich Ratzel and the Nature of (Political) Geography. In *Political Geography*, 19, 943–955.

Farinelli, F. (2003): *Geografia: Un'introduzione ai Modelli del Mondo*. Torino: Einaudi.

Frascani, P. (2008): *Il mare*. Bologna: Il Mulino.

Gambi, L. (1964): *Questioni di Geografia*. Napoli: Edizioni scientifiche italiane.

Gambi, L. (1971): Prefazione. In R. Mainardi (Ed.): *Le Grandi Città Italiane*. Milano: Franco Angeli, 3–4.

Gambi, L. (1973): *Una Geografia per la Storia*. Torino: Einaudi.

Gambi, L. (1994): Geography and Imperialism in Italy: Frome the "Unity" of the Nation to the "New" Roman Empire. In Godlewska, A.; Smith, N. (Eds.): *Geography and Empire*. Oxford/Cambridge: Blackwell, 74–91.

Giusti, U. (1938): Reazione Generale. In *Lo Spopolamento Montano in Italia: Indagine Geografico-economico-agraria*. Roma/Milano: Treves, volume 8.

Harvey, D. (1972): Revolutionary and Counter-Revolutionary Theory in Geography and the Problem of Ghetto Formation. In *Antipode*, 4, 1–13.

Jaja, G. (1936): *Il Porto di Genova*. Roma: Anonima Romana Editoriale.

Jureit, U. (2018): Mastering Space: Laws of Movement and the Grip on the Soil. In *Journal of Historical Geography*, 61, 81–85.

Koselleck, R. (1959): *Kritik und Krise: ein Beitrag zur Pathogenese der Bürgerliche Welt Sprache*. Freiburg: Alber.

Lando, F. (2020): *Per una Storia del Moderno Pensiero Geografico. Passaggi Significativi*. Milano: Franco Angeli.

Lefebvre, H. (1974): *La Production de L'espace*. Paris: Anthropos.

Marinelli, O. (1902): Alcune Questioni Relative al Moderno Indirizzo Della Geografia. In *Rivista Geografica Italiana*, 9, 217–240.

Marinelli, O. (1916a): La Geografia in Italia. Discorso di Olinto Marinelli con Alcune Appendici. In *Rivista Geografica Italiana*, 23(1–24), 113–131.

Marinelli, O. (1916b): I Tipi Ideali e la Descrizione delle Forme del Suolo. In *Rivista Geografica Italiana*, 23, 353–354.

Marinelli, O. (1916c): Dei Tipi Economici dei Centri Abitati. A Proposito di Alcune Città Italiane e Americane. In *Rivista Geografica Italiana*, 23, 413–431.

Martelli, M. (2001): La Geografia. In Simili, R.; Paoloni, G. (Eds.): *Per una Storia del Consiglio Nazionale Delle Ricerche*. Roma/Bari: Laterza, 492–509.

Nice, B. (1940): *La Casa Rurale Nella Venezia Giulia*. Bologna: Zanichelli.

Nice, B. (1946): Note Statistiche su la Distribuzione e la Popolazione delle Grandi città con Speciale Riguardo alla Marittimità e All'altimetria. In *Bollettino della Società Geografica Italiana*, 11, 6–15.

Nice, B. (1947): I Centri Abitati Della Toscana con Pianta Regolare. In *L'universo*, 27, 49–57.

Nice, B. (1952): La Pianificazione Territoriale Nello Sviluppo del Paesaggio Geografico. In Capello, C. F. (Ed.): *Atti del XV Congresso Geografico Italiano*. Torino: Iter, 532–536.

Nice, B. (1953): Geografia e Pianificazione Territoriale. In *Memorie di Geografia Economica*, 9, 7–153.

Pecout, G. (1997): *Naissance de l'Italie Contemporaine: 1770–1922*. Paris: Nathan.

Proto, M. (2014a): *I Confini d'Italia: Geografie Della Nazione dall'Unità alla Grande Guerra*. Bologna: BUP.

Proto, M. (2014b): Giovanni Marinelli (1846–1900) and Olinto Marinelli (1874–1926). In Lorimer, H.; Withers, C. W. J. (Eds.): *Geographers: Biobiblio-graphical Studies*. London/New York: Bloomsbury, volume 33, 69–105.

Proto, M. (2017): Irredenta on the Map: Cesare Battisti and Trentino-Alto Adige Cartographies. In *J-Reading-Journal of Research and Didactics in Geography*, 2, 85–94.

Rupke, N. (2011): Afterword Putting the Geography of Science in Its Place. In Livingstone, D. N.; Withers, C. W. J. (Eds.): *Geographies of Nineteenth-Century Science*. Chicago: University of Chicago Press.

Salvati, M. (1982): *Stato e Industria Nella Ricostruzione: Alle Origini del Potere Democristiano, 1944–1949*. Milano: Feltrinelli.

Sofia, F. (2020): Lucio Gambi. In *Dizionario Biografico Degli Italiani*. Roma: Treccani.

Sturani, M. L. (1998): "I Giusti Confini Dell'Italia". La Rappresentazione Cartografica Della Nazione. In *Contemporanea*, 3, 427–446.

Toschi, U. (1937): *Appunti di Geografia Politica*. Bari: Macri.

Toschi, U. (1939): Razza Ambiente Economia. In *Geopolitica*, 6, 12–24.

Toschi, U.; Brambilla, F. (1954): *La Determinazione dell'area di Influenza di Ivrea*. Ivrea: Olivetti.

Turco, A. (1980): I Modelli nei Paradigmi Della Geografia Italiana. In Corna-Pellegrini, G.; Brusa, C. (Eds.): *La Ricerca Geografica in Italia 1960–1980*. Varese: Ask, 865–880.

Vallega, A. (1980): La Regione: tra Cultura e Società. In Corna-Pellegrini, G.; Brusa, C. (Eds.): *La Ricerca Geografica in Italia 1960–1980*. Varese: Ask, 741–748.

Vlora, N. R. (1977): Gravità Universale e Pendolarità Operaia. In *Bollettino della Società Geografica Italiana*, 6, 165–208.

Wardenga, U.; et al. (2011): *Der Verband Deutscher Berufsgeographen 1950–1979. Eine Sozial-Geschichtliche Studie zur Frühphase des DVAG*. Leipzig: Leibniz-Institut für Länderkunde.

Zanetto, G. (1979): Il Potenziale: da Modello a Strumento. In *Rivista Geografica Italiana*, 86, 298–320.

12 The early years

William Bunge and *Theoretical Geography*

Trevor J. Barnes and Luke R. Bergmann

William Bunge's (1962, 1966a) monograph *Theoretical Geography* was the most sophisticated and evangelical of the early writings on geography's quantitative revolution. On the one hand, it provided the movement with an intellectually robust conceptual and philosophical foundation, including numerous worked through techniques and examples. On the other, it provided bible-thumping fervour and praise, exhorting unbelievers to leave behind their old ways of regionalism and to convert to the new geographical fundamentalism of quantification and theory with their promise of scientific salvation. Kevin Cox (2001, 71) later called Bunge's book, "perhaps the seminal text of the spatial-quantitative revolution", while Pierce Lewis (1973, 131) labelled it, "a minor classic". In 2001, a *Progress in Human Geography* forum on Bunge's book dispensed with the qualifier (Cox 2001; McMillan 2001; Bunge 2001). *Theoretical Geography* was simply a "classic in human geography".[1]

That designation represented a significant reversal of fortune for the volume. Bunge (1928–2013) at first could not find even a publisher for his book, the manuscript turned down by the University of Washington Press. Only after help from a small group of people who knew Bunge and valued his work, along with a significant financial contribution by his father, was a small press in Lund, Sweden, C. W. K. Gleerup, persuaded to print Bunge's monograph.[2] But even after it was published, almost no anglophone geographical journal paid it attention. As Bunge (1974, 485) later wrote, "*Theoretical Geography* was forced into foreign publication and was barely reviewed at all on this Continent".[3] Following Oscar Wilde, Bunge might have said, "the book was a great success, but the audience was a failure".

There are three different versions of the text that Bunge singly titled, *Theoretical Geography*. The first is his 1960 doctoral thesis completed at the Department of Geography, University of Washington, Seattle. The second is a 1962 version published by C. W. K. Gleerup. Its first five chapters were more or less the same as Bunge's thesis. The dissertation had a sixth chapter on geometry followed by a very short "Concluding Statement" of only a half-page. Bunge deleted both the latter in the 1962 published edition, replacing them by two chapters: one on "Experimental and theoretical central place" and another on "Distance, nearness and geometry". The third version, also

DOI: 10.4324/9781003122104-12

by Gleerup, was published in 1966 and likely the best known and most widely circulated. It was billed as the "second revised and enlarged edition" (Bunge 1966a, flyleaf). The 1962 version was augmented by the addition of two more chapters (adding 79 pages to the book's length).[4] The first was on "The meaning of spatial relations", the second on "Patterns of location". In his "Introduction to second edition", Bunge (1966a, xiii–xvii) provided a rationale for the changes: there was "not nearly enough physical geography" (Bunge 1966a, xiii) in the earlier edition; and there was insufficient recognition of how theoretical geography affected "mathematical and especially philosophical matters" (Bunge 1966a, xv). Even after the changes were made, however, there was still little physical geography and Bunge's claim to have made contributions to "mathematical" and "philosophical matters" was later strongly contested by Sack (1972).[5]

The purposes of this chapter are twofold. First, using archival sources and a biographical approach we follow how and where Bunge wrote *Theoretical Geography* and its revisions. We trace Bunge: from his early life in Wisconsin, where he was born and later attended graduate school at the University of Wisconsin, Madison; to the University of Washington, Seattle, where he wrote his doctoral dissertation; to the University of Iowa, Iowa City; and finally, to Wayne State University, Detroit, MI, where he revised his thesis for publication and then revised the revision producing the second edition. Second, we examine the content of *Theoretical Geography* and its revisions to understand why the work was so novel and innovative, becoming so influential in the discipline, a "classic in human geography".

Writing Theoretical Geography

Scattered throughout his published writing, Bunge provides a potted autobiography. Born in 1928, he grew up within a Lutheran German-American family in La Crosse, Wisconsin, along the east bank of the Mississippi River. In a pull-out biographical box from his second book, *Fitzgerald*, he wrote that "Bunge generations alternate between money-making and cause-serving" (Bunge 1971, 135). His father made money as a mortgage banker (and just as well for the fate of *Theoretical Geography*), but his son vigorously served many causes both political and intellectual.

Bunge attended the small private liberal arts Beloit College located at the extreme southern end of the State, majoring in political science (1946–1950). With the outbreak of the Korean War in 1950 and having just graduated, in November of that same year he was drafted into the US Fifth Army. He was deployed to Fort McCoy, Tomah, WI. The Camp served as it had during the Second World War as a training ground for new army recruits before being shipped overseas for active service. Set within the Camp was the Chemical, Biological and Radiological Wartime School. Ironically, given Bunge's later critical cartographic rendering of nuclear apocalypse, *The Nuclear War Atlas* (1988), he was assigned to "teach atomic war" at that School (Bunge 1988, xi).

During this same period, he enrolled in a geography Extension Course offered by the University of Wisconsin, Madison. Faculty from there would travel to various outposts of the university to give classes (Bunge 1988, xi). Fatefully, Bunge's class was taught, as he put it, by "the first professional geographer [I met] . . . in my life, . . . Richard Hartshorne" (Bunge 1988, xi). At that time, Hartshorne (1899–1991) was likely the most famous American geographer alive. He was the author of *The Nature of Geography* (1939) that meticulously explicated, rigorously justified, and genealogically fixed the discipline like no other English language volume had before it. Arguing for regional geography, Hartshorne's book contended that regions could be only described (not explained as in science) and treated as unique. Consequently, geography was an ideographic discipline. "Regional geography . . . is essentially a descriptive science concerned with the description and interpretation of unique cases", wrote Hartshorne (1939, 449).

Bunge may have accepted that methodological position in his first geography course (or knowing him, maybe not), but certainly within less than a decade he was publicly leading the charge against both it and Hartshorne.[6] *Theoretical Geography* was his response, subsequently provoking a 17-year tempestuous epistolary relationship between him and Hartshorne characterized by violent disagreement and on occasion from Bunge cruel and unjustified accusations (Barnes 2016).

Demobbed from the Army and fixated by geography Bunge enrolled as an MS student in geography at Science Hall, the University of Wisconsin, Madison. Graduating in 1956, he directly entered the Geography PhD programme. As we will see, that is when the problems began. In the acknowledgments to his PhD thesis that in the end he completed at the University of Washington, Bunge said it was initially Reid Bryson, a Professor of Meteorology at Wisconsin, who steered him towards mathematics. From Bryson he received "a prolonged research assistantship . . . [and] was given invaluable leeway and encouragement" (Bunge 1960, no page number). "Later", as Bunge (1960, no page number) continued, "Professor John Neese, of the Department of Zoology at the same institution, tutored me in the philosophy of science". It is not exactly clear how, but as a result of those influences, Bunge turned against Hartshorne and the ideographic regionalism he and the Department of Geography professed. Instead, Bunge's geography was to be mathematical, undergirded and justified by a rigorous philosophy of science. In a 1953 paper, Kurt Schaefer, a geographer at the University of Iowa, already had begun to work towards the same intellectual end, drawing on logical positivism to promote geography as a universal law-seeking (nomothetic) science. Moreover, Schaefer singled out Hartshorne's ideographic regionalism as both lacking intellectual warrant and as an obstacle that must be overturned for a true scientific geography to be practised. Schaefer's death before his paper was even published meant that he could not fight that fight, but Bunge could and did.[7]

Unsurprisingly, Bunge's subsequent doctoral comprehensive exams at Science Hall did not go well. He failed miserably, with Hartshorne casting a negative vote. Hartshorne later wrote to him, "in your prelims here you failed to show how much you did know of what you thought were less important things because you were dominated by the feeling that they were unimportant". In other words, Bunge refused to toe the ideographic regionalist line. In his sharp reply, Bunge made it clear that it was never a line that should have been drawn in the first place: "As for my flunk out, my crime . . . was not as serious as the Wisconsin's staff, stupidity" (Hartshorne 1959).

By the time Bunge received the results of his comprehensive exams, he already had plans to be a visiting student at the University of Washington, Seattle. He knew that some in that Department were carrying out the kind of "nascent mathematical geography" he wanted to pursue (Bergmann and Morrill 2018, 291). Once he arrived in Seattle, he was "allowed to stay [on] . . . for his doctorate" (Bergmann and Morrill 2018, 291). He became a member of the "space cadets", a name originally coined as a form of ridicule but later worn as a badge of honour by the group of quantitative and theoretically minded graduate students who had serendipitously gathered at the University of Washington from 1955 to 1958 to practise that "nascent mathematical geography".[8] In part they were guided by three youngish professors: the most well-known and senior was the former Harvard urbanist Edward Ullman (1941), who as a graduate student had introduced Walter Christaller's Central Place Theory to American geographers; second, was the analytical cartographer John Sherman, originally a climatologist and steeped in mathematics; and the most important, William Garrison, who acquired his mathematical skills as a meteorologist serving in the US Army Airforce in the Pacific Theatre during World War II and who acted as a mentor, muse, paymaster and protector for several of the cadets including Bunge whose thesis he also supervised.

At the heart of the Washington project was joining the traditional geographical interest in regionalism with the new interest in mathematizing, theoretically explaining and exactingly testing spatial relations. Jamming the two projects together, the old regional approach and the new law-seeking spatial theory, created the spark that ignited geography's quantitative and theoretical revolution (Barnes 2018a). That this happened in Seattle was a result of several contingent factors: a sympathetic Chair at the Department of Geography, Donald Hudson – "the midwife of the quantitative revolution"; the opportune but chance arrival from the mid-1950s of that group of young, ambitious, energetic, talented but intellectually frustrated, testosterone-charged male graduate students, "the space cadets"; Cold War research money especially from the Geography Section of the Office of Naval Research administered by Evelyn Pruitt (1979); interdisciplinary interactions within the university between geography and especially computational savvy faculty from Civil Engineering and Urban Planning (Bob Hennes and Edgar Horwood); the arrival in 1955 of the university's first computer, an IBM 604, housed in the attic of the Chemistry Building; and last but not least unlimited use of the Department's duplicating machine, indispensable for the production of intra-departmental and later internationally transmitted discussion papers (Barnes 2004).

For Bunge, perhaps the most important factor that transformed these ripe conditions for revolution into their creative realization was the culture of intellectual engagement among the "Cadets". Bunge crowingly describes that culture to Hartshorne in a letter in June 1959, rubbing it in:

> The arguments [here] . . . drag on month after month and often erupt into pure emotionalism but always return to the rational game. No one is sacred or has enough prestige to command respect. I think this is partly due to the extreme wealth of ideas and fierce work so that no one can really get ahead and also the youth of the staff. Here is an intellectual market place. . . . Students barge into offices, sharpshoot during lectures, quarrel openly and sometimes personally (although seldom feud) with staff. Here men [sic] sink or swim on their intellectual vigour and this is one of those rare places in the world or time, a department enjoying a golden period. It is such an atmospheric thing that I am always afraid something will upset its rare metabolism.
>
> (Hartshorne 1959)

It was not Science Hall.

Bunge's PhD thesis was likely the most glittering of the achievements of that "golden period". Bunge was going to call his dissertation, *Fundamental Geography* (Hartshorne 1959). It might have made a better title than *Theoretical Geography*, indicating the root and branch character of the disciplinary change Bunge envisaged. As Bunge's "Thesis Reading Committee" – William Garrison, Rhoads Murphy and Donald Hudson – put it in their 23 February 1960, commendatory examination reports: the dissertation's "stress on the theoretical and abstract . . . embraces all aspects of geography . . . [and] represents a significant contribution to knowledge" (Bunge 1960). Or as Bunge wrote in a 1959 swaggering letter to Hartshorne, "*We are achieving universality at the theoretical level. . . .* We are theoretical or fundamental geographers" (emphasis in the original) (Hartshorne 1959).

Completed, Bunge sent his thesis for publication to the University of Washington Press. In turn, they sent it to Hartshorne to review. He was not amused. According to Bunge, he "wrote [that] the book should be burned!" (Hägerstrand 1960)[9] Further, Hartshorne wrote a chiding letter to Donald Hudson requesting that his name be removed from the acknowledgements of the thesis. Bunge had thanked Hartshorne for his "close criticism" of his first chapter on "Method". That was beyond the pale for Hartshorne. He requested that his "name be omitted as unnecessary and misleading to the reader" because his actual comments on that chapter showed that Bunge "insist[ed] on the perpetuation of a fraud"[10] (Hartshorne 1960). Shut out of the University of Washington Press, Bunge contacted Torsten Hägerstrand who had visited and taught at the University of Washington, Spring Quarter, 1959. "I am so interested in seeing this book printed that I will do most anything, including publishing it myself, to see it born", Bunge wrote Hägerstrand (1960). That was not far short of what happened. Hägerstrand, believing that Bunge was wronged by Hartshorne, accepted the volume for his Lund Studies in Geography, Series C General and Mathematical Geography (Gould 1979, 141). It became the series first volume ("No. 1"). Bunge's father, and fortunately in the "money-making" half of the Bunge family, cut a cheque for US $1,800 for the small Lund publisher, C. W. K. Gleerup. The presses rolled, and 800 copies of the 1962 first edition of *Theoretical Geography* were printed (Hägerstrand 1960).

When the book went to press the original thesis was revised, made about a third larger. Bunge primarily made those revisions while he had a one-year sessional appointment at the University of Iowa (1960–1961). It was home to his martyred intellectual hero, Kurt Schaefer (1904–1953), who, like Bunge, also stood up to Hartshorne and criticized his ideographic position.[11] The founding Iowa Departmental Geography Chair, Harold McCarthy, had criticized Schaefer for attacking Hartshorne. He was still there when Bunge joined the Department and not keen on him either especially after he read the first chapter of *Theoretical Geography* that so exercised Hartshorne. In the Iowa student paper, *The Daily Iowan*, Bunge accused McCarthy of trying to muzzle him by preventing that chapter's publication (Hartshorne 1960).[12] Unsurprisingly, Bunge's contract was not renewed.

Such was the robustness of the American academic job market of the early 1960s, however, that despite Bunge's already muddied reputation and the power of Hartshorne within American geography, Bunge was immediately able to find another position. He moved to Detroit in late summer 1961 to begin at Wayne State University. At least early on in his tenure at Wayne State he continued to focus on the intellectual agenda of *Theoretical Geography* (Heynen and Barnes 2011). To that end, the same year he arrived in Detroit he helped to launch the Michigan Interuniversity Community of Mathematical Geographers (MICMOG). Meeting once a month in the back room of a tavern in Brighton, Michigan, the participants, several of them former space cadets, talked Greek letters and the language of geometry.[13] The next year he also began working with another fundamental geographer, Bill Warntz, at the American Geographical Society based in New York, to write "a successor" to *Theoretical Geography*. Titled *Geography: The Innocent Science*, it was an undergraduate primer for the new geography. It was equally as evangelical as *Theoretical Geography* and accompanied by seductive props that included an inflatable beach ball and a mini wooden Varignon frame. In the end, *Geography: The Innocent Science* was never published despite several drafts over many decades. But in 1966 a second "enlarged and revised" edition of *Theoretical Geography* was published. With geography's quantitative revolution now well under way, it became the most well-known edition, celebrated as a "classic" by *Progress in Human Geography* 35 years later.

Reading Theoretical Geography

Theoretical geography as Bunge used the term meant deploying a formal mathematical vocabulary to reduce complex spatial relationships to simpler geographical patterns that could then be theoretically explained. As an approach, it had a long history in geography. It was present at the very beginning of the subject in the West with the Ancient Greek (or Hellenised Egyptian) geographer cum astronomer, Claudius Ptolemy (63–181 CE) and his attempt to represent *geos*, the entire space of the world (Lukermann 1961). After a 1000-year gap or so, it was picked up in Enlightenment Europe by Bernhardus Varenius (1622–50), a young German geographer living in Amsterdam, forming the basis of his widely read and used *Geographia Generalis* (1650). Finally, it entered academic (university) geography during the late 19th and early 20th century in the form of degree courses in cartography, geodesy and surveying (Barnes 2011).

In mid-20th century, Bunge, along with the other cadets, their professors, as well as a scattering of others, were in effect trying to recoup this earlier tradition for post-war Anglophone geography. While there remained basic similarities between the new and old versions of the project, there were also differences. One was that in the new version of theoretical geography there were now many more mathematical concepts and techniques available, especially geometrical and statistical. For Bunge, it was vital that these new approaches be incorporated; the project was too important for them to be left out. On page 3 of his thesis, he quoted Moses Richardson's (1958) *Fundamentals of Mathematics* as justification:

> [M]athematics is basic to every subject forming part of the search for truth. . . . To mathematize a subject does not mean merely to introduce equations and formulas into it, but rather to mould and fuse it into a coherent whole, with its postulates and assumptions clearly recognized, its definitions faultlessly drawn, and its conclusions scrupulously exact.
>
> (Richardson 1958, 481)

One of Bunge's purposes in *Theoretical Geography* was therefore to review and apply relevant mathematical techniques. Half of the six substantive chapters in his thesis were devoted to that task: chapter 3, "A measure of shape", chapter 4, "Descriptive mathematics", and chapter 6, "Geometry applied to advanced theoretical geography". They enabled the older project of Ptolemy and Varenius to be extended and updated, refashioned for new times, ensuring that the current discipline, which for Bunge had been watered down by Hartshornian ideographic regionalism, to become again logically "coherent", "scrupulously exact" and adequately equipped for the "search for truth".[14]

A second difference was that there were now an increasing number of theories and principles that could be drawn upon to explain mathematized spatial patterns. Bunge's thesis reviewed them in chapter 5, "Advanced theoretical geography". It set out some of the theoretical options and their potential uses by geographers. Bunge further believed that geographers did not have to invent their own theory still to be theorists. They could also beg, borrow and steal theory developed in other fields. As he wrote, "once theory is produced it often can be applied to a variety of subjects . . . [demonstrating] a unity to knowledge" (Bunge 1960, 7). The only important issue for Bunge was whether the mathematical relations embedded within the appropriated theory matched the mathematical relations that described the geographical phenomenon. If it did, then it was quite possible that "a theory . . . constructed for climatology-oceanography . . . [could be] applied to population geography and economic geography, [or] a theory designed to explain the distribution of heavenly bodies . . . [could be] applied to contagious diseases, the movement of human ideas, and human and animal migrations" (Bunge 1960, 23). All that was needed was an isomorphism between "the implicit abstract logic of one theory with another" (Bunge 1960, 8). "All is fair in theory construction", Bunge (1960, 8) claimed.

That said, Bunge came increasingly to admire especially the German geographer Walter Christaller (1899–1968), who while drawing on some pre-existing elements had synthesized an original geographical theory, central place theory. In both the first and second editions of the published versions of *Theoretical Geography*, Bunge devoted a completely new chapter to Christaller's central place theory: "chapter 6, Experiments and Theoretical Geography". Indeed, on the flyleaves of the published versions, but not in the thesis, Bunge prominently dedicated *Theoretical Geography* "to Walter Christaller".[15] Bunge's "good wife, Betty", in contrast, received only a single sentence in the acknowledgments for having "stood by so steadfastly" (Bunge 1962, 1966a, ix).

Yet a third difference was that the new version of theoretical geography was justified by a larger philosophical rationale. That went to Bunge's interest in, sometimes obsession with, Kurt Schaefer and his disagreement with Richard Hartshorne. Bunge philosophically justified *Theoretical Geography* in the first chapter – "Methodology" in the thesis, "A Geographic Methodology" in the published versions – by drawing on Schaefer's (1953) *Annals* paper based on the philosophy of positivism, and in particular, the 20th-century interwar version known as logical positivism. In turn, Schaefer was influenced by a colleague at Iowa, the Austrian émigré philosopher of science, Carl Bergmann (1905–1987).[16] As a graduate student at the University of Vienna during the late 1920s, Bergmann was a member of the Vienna Circle that developed logical positivism. It said that the only real knowledge was scientific knowledge. All other knowledge was unreliable or spurious. Scientific knowledge for logical positivism was knowledge capable of expression as a universal law taking the logical form: if cause A, then everywhere and for all time the same effect, B. Schaefer argued in his paper, which Bunge (1960, 13–20, 22–29) accepted, that if geography's basic spatial unit, the region, was taken as unique – Hartshorne's position in *The Nature of Geography* – then geography could never produce laws and thus could never produce scientific knowledge. As Schaefer (1953, 236) put it, "there are no laws for the unique". Hence, geography would be stuck forever with producing unreliable and spurious knowledge.[17]

Bunge's purpose in *Theoretical Geography* was to demonstrate that the discipline was a science. Hartshorne was wrong. Regions were not unique, geographical universal laws could be formulated. His argument was sketched out in chapter 1 of his thesis and further elaborated in expanded versions of the same chapter in the later published versions. He claimed that a region was nothing more than a form of areal classification based on a set empirical criteria (Bunge 1962, 1966a, 14–15). Wet regions are areas in which rainfall crosses a certain empirical threshold, or rich regions are areas in which incomes meet or exceed that numerical amount that defines richness. In this interpretation, a region is defined by facts, in these examples, rainfall or income. Moreover, "these facts are general rather than unique" (Bunge 1962, 1966, 14). All wet regions are empirically wet in the same way, just as all rich regions are empirically rich in the same way. With the region thus rendered uniform, Bunge argued general theoretical explanation is possible, as are the formulation of universal laws.

One further point: theoretical geography did not mean for Bunge the eradication of regional geography, only its redefinition. It would be the disciplinary branch that collected and stored general facts for filling out areal classification schemes with the ultimate purpose of serving geographical theory. Bunge believed regional geographers were perfectly positioned for that work. They had "mastered . . . the language"; they had "in hand . . . the locational frame of an area"; they knew "local customs"; and they were "familiar . . . [with] government facilities, personnel and regulations" (Bunge 1960, 22). If anyone could gather vigorous areal facts it was them. That task should never be conceived as an end in itself, though. That was Hartshorne's error. It must always be subservient to the higher purpose of theoretical geography and the generation of law-like generalizations.

One final difference was around the conception of the map. Ptolemy was the first person to use mathematics in cartography, in his case by devising a locational coordinate system for plotting a map of the known world. Mathematics was even more prominent in Varenius's (1650) *Geographia Generalis*, later revised and republished by Isaac Newton (Warntz 1989, 177). Bunge too thought cartography and mathematics were inextricably entwined and central to his project of theoretical geography. As Bunge (1960, 24) wrote in his doctoral thesis, "the reason geography has always paid such respect to maps is that they have been the logical framework upon which geographers have constructed geographic theory". The maps that littered *Theoretical Geography* were not the traditional kind, however, based on Euclidean geometry and conceptions of absolute space. In this sense, Bunge pushed the mathematics of cartography, drawing especially on the idea of map transformations that were contemporaneously developed at the University of Washington by another of the cadets, Waldo Tobler, and his doctoral supervisor, the cartographer John Sherman. Map transformations entailed redrawing maps using a different metric than the conventional measure of distance as miles or kilometres. For example, in one of his own first map transforms, Bunge (1960, figure 14 between pages 43 and 44) used time rather than distance to draw the commuter's map of Seattle. It radically altered cartographic space, becoming as O'Sullivan et al. (2018, 110) put it, a "more-than-Euclidean, non-absolute spatial representation". For Bunge, this kind of map demonstrated what an invigorated theoretical geography could accomplish. In contrast, maps of the "map thumpers" of the sort that populated Hartshorne's regional geography remained stiflingly unchanged since academic geography began in the mid-to-late 19th century (Bunge 1968a, 2). They required revolutionizing, further mathematizing, which map transformations achieved.

Conclusion

Bunge's *Theoretical Geography* was a radical experiment. Like all experiments, it involved: innovatively carrying out acts never tried before, such as generalizing cartographic distortion by drawing a map of Italy using only straight lines (Bunge 1960, figure 3 between pages 33 and 34), or testing the "shifting rule" by tracing similarities between a map of the changing course of the Mississippi River and a map of US State Highway 99 wending through Tacoma-Seattle (Bunge 1962, 1966a, 29–30); inventing a fresh vocabulary to describe original lines of inquiry, procedures and results, such as "metacartography" (Bunge 1966a, chap. 2), "the U graph", the "fundamental region" (Bunge 1966a, 229), "similitude transformation" (Bunge 1966a, 219), "morphological location" (Bunge 1966a, 257), and more; and trying to convince the larger community of the importance of the novel approach – in Bunge's case by producing three different versions of *Theoretical Geography*.

Bunge struggled especially with the last of these tasks. Partly this was because of his personality and the biographical baggage he brought to his project. Partly it was because he was the first person to set out a systematic monograph-length account and justification of what was a strikingly different kind of academic geography. His creative novelty was too strange, too out of sync with the then standard beliefs and practices of especially North American geographers, and similarly often out of sync with the projects he is reread to be in support of today (see Bergmann and Morrill 2018). How could the punctuated Adriatic coastline of Italy ever be cartographically rendered by straight lines? How could the naturally forming meanders of a river ever be like the human-constructed meanders of a highway? A comparison here is with Peter Haggett's (1965) monograph-length presentation also of the new geography, *Locational Analysis of Human Geography*. Haggett's book, in contrast, was a runaway success from the start, influencing human geography immediately and over the longer term (Barnes 2018b). But it came out five years after Bunge's dissertation was written and three years after the first edition of *Theoretical Geography* was published. By the time Haggett published his book, the ground for its success was already prepared and in part by Bunge. Not so for Bunge's own volume, where the ground was barely turned over. That's why it took over 40 years for Bunge's work to be recognized as a classic, and now over 60 years before it receives its own chapter in a volume about the history of geography's quantitative revolution. History is littered with people whose pioneering works and ideas were not recognized by their own communities within their own time. William Bunge was one of them. He is geography's Franz Kafka, Vincent van Gogh and John Keats.

Notes

1 Michael Goodchild (2008) also wrote a chapter celebrating *Theoretical Geography* for Hubbard, Kitchen and Valentine's important edited book, *Key Texts in Human Geography*, a collection that reviewed the most important English-language monographs published by geographers over the preceding 50 years.

2 C. W. K. Gleerup became an important publisher of primarily Swedish geographical research monographs with a close link to the Department of Geography, Lund University. The volumes were published as Lund Studies in Geography. Bunge's (1962, flyleaf) *Theoretical Geography* inaugurated the Lund Studies new "Series C, Mathematical Geography". A partial listing of volumes within Lund Studies is found at the back of Bunge's (1966a) volume.

3 The interest in Bunge's book in the non-anglophone world is shown by the often relatively quick translation and publication of *Theoretical Geography* outside of English-speaking geography. Bergmann and Morrill (2018, 297) note that "*Theoretical Geography* has been translated into, among others, Russian (1967), Japanese (1969), Spanish (in part, 1969), Chinese (1991), and French (see Popelard et al. 2009)".

4 The two extra chapters were also published by Gleerup as a standalone publication (Bunge 1966b).

5 The issue turned on Bunge's (1966a, 200) claim that "geometry and [geographical] movement are inseparable duals". Sack (1972, 72) argued that could not be the case because there is no time or change within the axioms of geometry. Without them, geographical movement is ruled out by definition. Geometry and movement are therefore mutually exclusive. They can never be duals of one another.

6 In his correspondence with Hartshorne, Bunge wildly swung from deference and praise to searing critique and contemptuous dismissal (Barnes 2016). By the late 1960s, however, he realized that he had been wrong to dismiss Hartshornian regionalism. In 1976 in an interview with Don Janelle (1976) he said, "Yes, I have been forced to admit I was wrong on that [i.e., regionalism], which is very painful; it's the only thing I've ever been wrong at, and I'll never be wrong again, so don't think it's habitual." Bunge's work became increasingly additive, trying to join seemingly opposed positions, like the ideographical and the nomothetic, regionalism and spatial science (see Bunge 1974). Bringing together ostensibly opposed ideas became for him a source of creativity (Barnes 2018a).

7 Schaefer (1953) died of a heart attack at a matinee show in an Iowa City cinema on 3 June 1953, before his paper was published in the September 1953 issue of *Annals of the Association of American Geographers*. See also fn. 11.

8 The original space cadets were Brian Berry, Ronald Boyce, Duane Marble, Richard Morrill and John Nystuen. William Bunge, Michael Dacey, Arthur Getis and Waldo Tobler later joined them. Joe Spencer, an Asian specialist at UCLA, coined the name that was meant to be facetious to describe the group at the 1956 Seattle Pacific Coast regional AAG meeting.

9 William Bunge wrote to Torsten Hägerstrand, 22 June 1960.

10 The "fraud" was Schaefer's 1953 criticism of Hartshorne. Hartshorne believed that criticism rested on fraudulent use of sources and anyone who wrote in favour of Schaefer, such as Bunge in his thesis, continued the deception.

11 Martyred because Bunge (1968b, 7–8) in an infamous paper, "Fred K. Schaefer and the science of geography", implied Schaefer, who was a communist, died from hounding by the FBI. Bunge also suggested that investigation was triggered by Hartshorne who during the Second World War was a senior member of the Office of Strategic Services, forerunner of the CIA. Hartshorne tried to have Bunge's paper removed from any library that held it, although he was unsuccessful. Hartshorne adamantly denied involvement in any FBI surveillance of Schaefer and was skeptical that any took place: "No FBI man ever asked me about Schaefer, and if they had I would have said: 'I don't know. I have no reason to suppose he was not an enthusiastic American citizen'" (Dow 1986, 5).

12 "4 geography Profs speak up on Bunge", *The Daily Iowan*, 21 November 1960. (Hartshorne 1960)
13 The MICMOG discussion paper series is available online: http://deepblue.lib.umich.edu/handle/2027.42/58252
14 Ironically, Hartshorne's undergraduate degree was in mathematics from Princeton.
15 Bunge (1977) continued to defend and support Christaller even after news leaked out that he worked for the SS during the Second World War and was a member of the Nazi Party.
16 There were several similarities between the two men, connecting them. Both were of a similar age (Schaefer was a year older); both were native German speakers; both were forced to flee their native countries as political refugees and come to America because of the Nazis – in 1933 Schaefer left Germany because of his left-wing political views, and in 1938 Bergmann left Austria following the *Anschluss* because he was Jewish; both started to teach at the University of Iowa the same year, 1939; and both were unreservedly champions of science and the scientific method.
17 Exactly that same judgment was made in 1948 by James Conant, President of Harvard, when he closed his university's Department of Geography. He believed that geography was incapable of producing scientific knowledge of the type he knew as a former Harvard Professor of Chemistry. As a result, geography was "not a university subject" (quoted in Smith 1987, 159).

References

Barnes, T. J. (2004): Placing Ideas: *Genius Loci*, Heterotopia, and Geography's Quantitative Revolution. In *Progress in Human Geography*, 29, 565–595.

Barnes, T. J. (2011): Spatial Analysis. In Agnew, J.; Livingstone, D. L. (Eds.): *The Sage Handbook of Geographical Knowledge*. London: SAGE, 380–391.

Barnes, T. J. (2016): The Odd Couple: Richard Hartshorne and William Bunge. In *The Canadian Geographer*, 60, 459–465.

Barnes, T. J. (2018a): A Marginal Man and His Central Contributions: The Creative Spaces of William ("Wild Bill") Bunge and American Geography. In *Environment and Planning A*, 50, 1697–1715.

Barnes, T. J. (2018b): A Hundred-Year Classic: Peter Haggett's *Locational Analysis in Human Geography* (1965). In *Geografiska Annaler*, 100, 294–297.

Barnes, T. J.; Farish, M. (2006): Between Regions: Science, Militarism, and American Geography from World War to Cold War. In *Annals, Association of American Geographers*, 96, 807–826.

Bergmann, L.; Morrill, R. (2018): William Wheeler Bunge: Radical Geographer (1928–2013). In *Annals of the Association of American Geographers*, 108, 291–300.

Bunge, W. (1960): *Theoretical Geography*. Ph.D. dissertation. Seattle: Department of Geography, University of Washington.

Bunge, W. (1962): *Theoretical Geography*. Lund: C W K Gleerup.

Bunge, W. (1966a): *Theoretical Geography*. Lund: C W K Gleerup, 2nd revised and enlarged edition.

Bunge, W. (1966b): *Appendix to Theoretical Geography*. Lund: C W K Gleerup.

Bunge, W. (1968a): The Philosophy of Maps. In Bunge, W. (Ed.): *Michigan Inter-University Community of Mathematical Geographers*, Discussion Paper number 12, The Philosophy of maps.

Bunge, W. (1968b): Fred Schaefer and the Science of Geography. In *Harvard Papers in Theoretical Geography, Special Papers Series*, Paper A.

Bunge, W. (1971): *Fitzgerald: Geography of a Revolution*. Cambridge, MA: Schenkman.

Bunge, W. (1974): Regions Are Sort of Unique. In *Area*, 6, 92–99.

Bunge, W. (1977): Walter Christaller Was Not a Fascist. In *Ontario Geography*, 37, 84–86.

Bunge, W. (1988): *Nuclear War Atlas*. Oxford: Basil Blackwell.

Bunge, W. (2001): Classics in Human Geography Revisited: Bunge, W., *Theoretical Geography*. Author's response. In *Progress in Human Geography*, 25, 75–77.

Cox, K. R. (2001): Classics in Human Geography Revisited: Bunge, W., *Theoretical Geography*. Commentary 1. In *Progress in Human Geography*, 25.

Dow, M. W. (1986): Richard Hartshorne Interviewed by Maynard Western Dow, Minneapolis, MN, May 6. *Online Geographers on Film Transcriptions Geographers on Film Collection*, 12. Available at: extarchive.ru/c-2977382-pall.html (last accessed 23 January 2020).

Goodchild, M. (2008): William Bunge's *Theoretical Geography*. In Hubbard, P.; Kitchen, R.; Valentine, G. (Eds.): *Key Texts in Human Geography*. London: SAGE, 9–16.

Gould, P. (1979): Geography 1957–1977: The Augean Period. In *Annals of the Association of American Geographers*, 69, 139–151.

Hägerstrand, T. (1960): *Posthumous Papers of Torsten Hägerstrand*. Lund: Lund University, volume 37.

Haggett, P. (1965): *Locational Analysis in Human Geography*. London: Edward Arnold.

Hartshorne, R. (1939): *The Nature of Geography: A Critical Survey of Current Thought in the Light of the Past*. Lancaster, PA: Association of American Geographers.

Hartshorne, R. (1959): *Papers of Richard Hartshorne*. Box 194. Hartshorne correspondence – William Bunge, File F, American Geographical Society Library. University of Wisconsin, Milwaukee, WI.

Hartshorne, R. (1960): *Papers of Richard Hartshorne*. Box 194. Hartshorne correspondence – William Bunge, File F, American Geographical Society Library. University of Wisconsin, Milwaukee, WI.

Heynen, N.; Barnes, T. J. (2011): *Fitzgerald: Then and Now: New Introduction to Fitzgerald: Geography of a Revolution by William Bunge*. Athens, GA: University of Georgia Press, vii–xvi.

Janelle, D. (1976): William Bunge Interviewed by Don Janelle, London, ON, November 3. *Online Geographers on Film Transcriptions Geographers on Film Collection*, 6pp. Available at: extarchive.ru/c-2977382-pall.html (last accessed 23 January 2020).

Lewis, P. (1973): Review. In *Annals of the Association of American Geographers*, 63, 131–132.

Lukermann, F. E. (1961): The Concept of Location in Classical Geography. In *Annals of the Association of American Geographers*, 51, 194–210.

McMillan, W. (2001): Classics in Human Geography Revisited: Bunge, W., Theoretical Geography. Commentary 1. In *Progress in Human Geography*, 25, 73–75.

O'Sullivan, D.; Bergmann, L.; Thatcher, J. (2018): Spatiality, Maps and Mathematics in Critical Human Geography: Towards a Repetition with Difference. In *The Professional Geographer*, 70, 29–39.

Popelard, A.; Elie, G.; Vannier, P. (2009): William Bunge, le Géographe Révolutionnaire de Detroit [William Bunge, the revolutionary geographer of Detroit]. *Le Monde Diplomatique*, 29 December (accessed 9 March 2020).

Pruitt, E. L. (1979): The Office of Naval Research and Geography. In *Annals, Association of American Geographers*, 69, 103–108.

Richardson, M. (1958): *Fundamentals of Mathematics*. New York: MacMillan.

Sack, R. D. (1972): Geography, Geometry and Explanation. In *Annals of the Association of American Geographers*, 62, 61–78.

Schaefer, K. (1953): Exceptionalism in Geography: A Methodological Examination. In *Annals of the Association of American Geographers*, 43, 226–245.

Smith, N. (1987): "Academic War Over the Field of Geography": The Elimination of Geography at Harvard, 1947–1951. In *Annals of the Association of American Geographers*, 77, 155–172.
Ullman, E. L. (1941): A Theory of Location for Cities. In *American Journal of Sociology*, 46, 853–864.
Varenius, B. (1650) *Geographia Generalis*. Amsterdam: Louis Elzevir.
Warntz, W. (1989): Newton, the Newtonians, and the *Geographia Generalis Varenii*. In *Annals of the Association of American Geographers*, 79, 165–169.

13 Mathematics against technocracy

Peter Gould and Alain Badiou

Matthew Hannah

The main purpose of this chapter is to draw attention to a particular strand of work by Peter Gould, who gained prominence as a practitioner and a vocal advocate of the so-called quantitative revolution of the 1960s and 1970s in Anglophone human geography. As Trevor Barnes has documented in a range of studies, the advent of more sophisticated quantitative methods among Anglophone geographers was a conflictual process rooted in very specific, if complex networks of people, places, and practices (Barnes 2001, 2004). Other chapters in the present volume show just how differently shifts toward quantitative methods were inflected by other people, spaces, and places in other parts of the world. The Anglophone quantitative revolution was shaped to an important extent by the experiences of US and British geographers in World War II (Barnes 2004; Barnes and Farish 2006). Other contextual factors also played a role, such as the disciplinary insecurity prompted by the closure of Geography at the prestigious Harvard University (Smith 1987). As elsewhere, attempts to assert the legitimacy of a focus on statistical data and advanced statistical methods and models met with resistance and led to lasting antagonisms.

Peter Gould, who completed his PhD with Ned Taaffe at Northwestern University, maintained close professional and personal ties with many of the most prominent champions of the new paradigm, such as David Harvey, Torsten Hägerstrand, Gunnar Olsson, and Peter Haggett. He was known not only as a creative and original user of quantitative methods but also as one of the sharpest critics of the more traditional strands of regional geography against which he and his allies struggled. His oft-cited retrospective 1979 paper on the process, titled "Geography: the Augean period," likened the knowledge produced by traditional geography to the manure in the mythical Augean stables, and the quantifiers to a collective Hercules tasked with clearing it out (Gould 1979).

Gould's overall corpus was wide-ranging in thematic terms. His best known research monograph is probably *Mental Maps*, co-authored with Rodney White (Gould and White 2012). Yet he also published spatial analyses of soccer games (Gould and Gatrell 1980), structures and flows of international television programming (Gould and Johnson 1978, 1980), the geographical and democratic aftermath of the Chernobyl reactor explosion (Gould 1990), and the spread of HIV-AIDS (Gould 1993), among other topics. His textbook on methods of

DOI: 10.4324/9781003122104-13

spatial analysis for university geography students, titled *Spatial Organization* and co-authored with Ron Abler and John Adams, was known as the "Blue Bible," and was widely used in US geography departments during the 1970s and 1980s (Abler et al. 1971). In addition, Gould continued to pen provocative and contentious reflections on the development of the discipline of human geography (Gould 1985, 2000). Partly for this reason, he became a *persona non grata* among many critical human geographers, and his work has not garnered much comment since his death in 2000 (though see Wyly 2014, 33). For full disclosure, I should say here that Gould was the supervisor of both my MSc and PhD theses at Penn State between 1986 and 1992. But the purpose of this chapter is not any kind of "rehabilitation" motivated by loyalty. I found and still find many of Gould's writings on developments in human geography highly problematic or flat wrong. Nevertheless, some of his ideas about the possibilities inherent in quantitative methods are original enough to deserve more consideration than they have been given up to now.

Like a number of his contemporaries, most notably, David Harvey, Doreen Massey, Gunnar Olsson, and Allen Pred, Gould turned in the 1970s to philosophical and theoretical perspectives that challenged the positivist-scientistic worldview and the social role of the forms of spatial analysis with which he had made his name. The development of historical-geographical materialism undertaken in different ways by Harvey and Massey has of course been extremely important. Allen Pred's and especially Gunnar Olsson's more idiosyncratic and linguistically oriented trajectories qualify them in retrospect as creative pioneers of post-structuralism in human geography. The project Gould embarked upon in the 1970s would not garner as much attention or influence, in part because of his habit of picking fights and making enemies within the discipline. Nevertheless, the postpositivist project Gould undertook was in one sense more ambitious than that of his contemporaries. Gould engaged deeply with the work of Martin Heidegger and became a self-described Heideggerian. At the same time, more than Pred, Harvey, and even Olsson, Gould wanted to defend the value and to creatively expand the possibilities of representing the human world mathematically. As Stuart Elden's *Speaking Against Number* makes clear, reconciling these two commitments is necessarily a difficult proposition to say the least (Elden 2006).

This paper is an attempt to frame and explain the way Gould approached the problem of reconciling Heidegger with spatial analysis. Instead of giving a comprehensive account, the argument focuses upon two important articles from 1980 and 1981. The bulk of the chapter is devoted to contextualizing and explaining Gould's arguments. The latter part of the chapter links them tentatively with Alain Badiou's ongoing project of rescuing mathematical thinking from the negative connotations it has acquired in the critical cultural and social thought of recent decades (Badiou 2006, 2008, 2009). Badiou's "phenomenological" enterprise shows some striking similarities to Gould's, but Badiou raises additional possibilities not foreseen by Gould for vindicating a mathematical ontology. By setting these two projects in sequence, we can get a sense of some core issues facing recent geographical work seeking to refound spatial analysis on a post-technocratic, critical footing (Hannah 2001; Johnston et al. 2014; Keylock and Dorling 2004; Wyly 2009, 2014).

"Letting the data speak for themselves"

Starting in the 1970s, Peter Gould began to read Heidegger's work under the tutelage of Joseph Kockelmans, who was Professor of Philosophy at Pennsylvania State University and one of the foremost Heidegger scholars of the postwar period. At Penn State, Kockelmans presided over a long-running, interdisciplinary philosophy seminar in which Gould was heavily involved (Gould in fact financed English translations of some pieces read by the group). Gould took a number of things from his long and multifaceted engagement with Heidegger and other continental thinkers, first of all an appreciation for the general phenomenological commitment to careful observation of that which appears to us, with a minimum of unreflected preconceptions. This is not specifically Heideggerian, of course, but was an ethos embodied in the practice of reading Kockelmans espoused. One more specifically Heideggerian point Gould found compelling was the *Dasein*-analytic alternative to Freudian psychoanalysis. Gould nurtured a strong animosity toward psychoanalysis and frequently invoked the work of Heideggerian existential psychologists like Medard Boss as superior to Freudian psychotherapy. Most important to Gould's attempted reconciliation of mathematical geographical methods with Heidegger was the latter's creative rereading of the ancient Greek notion of truth as "aletheia" or the unconcealment of that which presences.

Despite his long and varied engagement, Gould did not write in depth or at length about Heidegger, perhaps because he was keenly aware that in Kockelmans he had a world expert looking over his shoulder. However, he did produce a few articles that ventured into more detail. The most important of these appeared in 1981 in the *Annals of the Association of American Geographers* under the title, "Letting the data speak for themselves" (Gould 1981).[1] The title already gives a sense for how the reconciliation between a Heideggerian sensibility and mathematical representation is supposed to work: as the application of a phenomenological ethos of careful, presupposition-free observation to (at least potentially) quantitative data.

Gould's chief philosophical reference for this paper is Heidegger's essay "Science and Reflection," based on a 1954 lecture he gave in Munich (Heidegger 1977). In this essay, Heidegger seeks to characterize modern science as "the theory of the real" (*Ibid.*, 157). The bulk of the essay is taken up by explication of what is meant by the two terms "theory" and "the real." In both cases, Heidegger begins with their meanings in ancient Greek, and then traces the ways in which these original meanings were transformed, first by the Romans, and then in their modern incarnations. The "real," in the original Greek sense of "the presencing, consummated in itself, of self-bringing-forth," becomes for the Romans the result of an *operatio*: the real becomes a mere "consequence" of something done (*Ibid.*, 160, 161). Similarly, "theory," which Heidegger reads in terms of truth as "aletheia," meant for the Greeks the "reverent paying heed to the unconcealment of what presences." In the hands of the Romans, however, to theorize becomes *contemplari*, "to partition something off into a separate sector and enclose it therein" (*Ibid.*, 164, 165). Modern science then, effectively continuing the Roman lineage, "sets upon" the real in the "objectness" it has been given through "templating" or partitioning (*Ibid.*, 167). For Heidegger, however, the story is not simply one of degeneration: vestiges of the original Greek understanding survive inconspicuously even in the modern forms and practices of science. The task, then, is to recover these original dimensions (*Ibid.*, 171ff).

The survival of a richer, "Greek" understanding of reality and of theory, even in the driest procedures of advanced modern science, is the Heideggerian idea around which Gould organizes his 1981 essay. At the beginning of "Letting the data speak for themselves," Gould playfully acknowledges that the title appears naïve, admitting that "inanimate data can never speak for themselves" (Gould 1981, 166). But he immediately suggests that "there may be a deep vein of truth underneath such provocation" (Gould 1981). The early sections of the essay are dedicated to showing how often accepted models of reality have been discarded for better understandings (examples are taken from physics and – especially Freudian – psychology). This discussion serves as a vehicle for introducing the phenomenological commitment to observation of what presents itself (Gould 1981, 168–171). Gould then introduces Heidegger more explicitly as advocating "an attitude of questioning, that, no matter how much distorted by subsequent translations into Latin and other European tongues, can still come through to us if we are prepared to listen" (Gould 1981, 172). Citing "Science and Reflection," Gould describes the desired attitude in terms of Heidegger's definition of theory, a "reverent paying heed to the phenomena themselves" (Gould 1981).

At this point Gould turns to quantitative methods in human geography, and makes three general critiques of spatial analysis as practiced up to that time. First, in borrowing modes of description from the physical and biological sciences to characterize human phenomena, geographers have "objectified, cut off, 'templated' the very beings-in-the-world that should be of our deepest concern." Second, geographers have relied too readily upon simplistic cartographic and algebraic languages that "not only limit, but actually crush out of existence that which might be open to our paying heed." "Crushing" he understands above all as the reduction in dimensionality that occurs when human life is projected onto simplifying grids, as in factor or cluster analysis. Third, according to Gould, geographers have relied too heavily upon a priori models, and have been misguided in investing so much effort to fitting data to these models (Gould 1981, 174).

If one accepts Gould's initial critique, what would then be an alternative approach? Gould readily admits that some sort of interpretive framework is unavoidable. "We are obliged, in any act of inquiry, to project the complexity of human life onto some framework, even some geometry, but such frameworks of meaning must be derived by phenomenological and descriptive methods" (Gould 1981). It is at this point in his argument that Gould most baldly expresses his continued commitment to mathematical representation, asserting that it is "doubtful that we can write meaningful empirical explanations without the use of mathematics" (Gould 1981, 174–175). Gould's lack of acknowledgment of the rich range of rigorous qualitative methods being developed in feminist geography and elsewhere at that time is consistent with the disparaging attitude he would display elsewhere toward a range of critical and emancipatory geographies (Gould 2000). This blind spot regarding qualitative methods implies that his intellectual project is not entirely independent of his rather reactionary disciplinary politics. Yet even if he is wrong to assert that "meaningful empirical explanations" must involve mathematics, in my view the line of thinking he develops on the basis of this error produces interesting insights.

To return to the 1981 paper, Gould wants to let the data speak for themselves, but does not want to disqualify quantitative information from the outset. He thus argues that it is "vital to choose a form of mathematics that does not violate that which has been heeded" (Gould 1981, 174). His candidate for a more appropriate mathematical language, briefly sketched in the last section of the 1981 paper, is Q-Analysis or polyhedral dynamics, as developed in the 1970s in a series of papers and books by the British mathematician Ronald Atkin (Gould 1981, 175–176). In "Letting the data speak for themselves," Gould leaves himself no room for an extended explanation of Q-analysis, sketching its most general features in a page of text. For somewhat more detail, it is necessary to consult the second of the two papers discussed here, which appeared in 1980 in the *International Journal of Man-Machine Studies*. Titled "Q-analysis, or a language of structure: an introduction for social scientists, geographers and planners," this paper lays out the basic principles of Q-analysis and gives a range of examples of how it has been used (Gould 1980). It is important to note before getting into the particulars that Atkins' Q-analysis was by no means the only such initiative at the time. A range of very similar techniques that came to be known as "social network analysis" had been under development by sociologists since the 1950s (Scott 1987). However, Gould does not cite this work, drawing instead only upon extensive reading of the work of Atkin and his student Johnson, with whom Gould himself published analyses of the problem of classifying television programs (Gould and Johnson 1978, 1980).

A topology of connectivity

As the title of the second paper already suggests, Q-analysis is centrally concerned with defining "structures" of many different kinds in ways that respect the multidimensionality of the phenomena in question. Multidimensionality is, again, the chief way in which Gould understands the complexity and richness of human social life. To capture this richness, so Gould argues, words are often inadequate, and while graphics of various sorts might capture simple structural relations, it is often necessary to resort to more abstract algebras capable of accommodating higher levels of multidimensionality beyond our ability to visualize (Gould 1981, 170). The operations typical of regression, factor and cluster analysis, in all of which complex relations are reduced to summary lines or neatly partitioned categories, literally collapse dimensions, and thus represent an understanding of structure Gould wishes to avoid.

Gould begins his explication of Q-analysis with a review of basic set theory, because the operations involved in any algebra are valid only in relation to carefully defined sets of data (Gould 1981, 173). As soon as we try to define aspects of the social or natural world in terms of sets, though, we see that relations between sets are often hierarchical but not exclusive. They are hierarchical in the sense that some sets "cover" or include others at lower levels of generality. But this covering is not exclusive, that is, it is seldom the case that each set at a given level is only included in one set at the next highest level. Gould illustrates these principles with the set of elements of a visual scene or landscape, in which, for example, a dandelion "belongs to . . . the words {weeds}, {flowers} and {vegetables} all at the same time" (Gould 1981, 174). A dandelion cannot be neatly partitioned into only one subset of the landscape. Put differently, a dandelion is a multidimensional phenomenon because it is a member of more than one category at more than one level.

Q-analysis seeks to describe as accurately as possible such hierarchical and nonexclusive relations between sets of data. On the basis of an initial description, the idea is to identify the structures of connectivity or lack of connectivity inherent in these relations. The term "relations" is key, because it is a much more general category than that of functions. A function, which prescribes strict – often linear – relations of dependency between two sets of variables, is a special case of a "mapping," which more broadly "relates the elements of one set to another" in some way (Gould 1981, 176–177). A mapping is in turn a subset of "relations" in the broadest sense. Thinking in these wider relational terms, Gould argues, allows us to escape the straitjacket of functional thinking, which he considers "the most constrained sort of thinking we can possibly employ" (Gould 1981, 177). "When we force our data into the form of a function, as in regression analysis and virtually all least-square multivariate techniques, we are actually crushing information out of our data sets and destroying all chance of recovering it again" (Gould 1981, 179).

To illustrate how Q-analysis avoids such reduction, Gould takes the example of relations between a hypothetical set of academics, perhaps members of a department, and the set of intellectual interests shared by at least some of these individuals. Represented as an incidence matrix, the relation between these two sets is a table in which, for each person (each row), a "1" is entered in each column representing an intellectual interest they have. In all other cells of the table, a "0" is entered. What stands out immediately is that there is no simple, 1:1 correspondence between individuals and interests. Some individuals have more interests than others, and some interests are shared by more individuals than others. Q-analysis represents these complex relations as polyhedra of different dimensionalities. A person who has four interests from among those surveyed would be represented by a tetrahedron with four vertices (interests), while someone with only two of the surveyed interests would appear as a line connecting two points. Conversely, the polyhedra could represent the *interests*, having higher or lower dimensionality according to how many or few vertices (this time, people) share them. In the language of Q-analysis, each basic unit, whether an individual or an interest, is called a "simplex" (Gould 1981, 180).

To stick with the first case, in which polyhedra represent people, there is also the question of their connectivity with others. It is not necessarily the case that someone with more interests is more connected. In fact, someone with many different interests might nevertheless be relatively disconnected from others, if those interests are unusual or idiosyncratic enough, not shared by others. By the same token, a person having only two interests might nevertheless be the sole link between two groups, each of which have many more interests. Groups of differentially linked polyhedra are called "simplicial complexes" (Gould 1981). The exact configurations of simplices within simplicial complexes are infinitely variable. And there is no upper limit to how many "dimensions" a unit can have, even if the graphical representation of polyhedra ceases to be useful beyond three or at most four dimensions. Additionally, connections between vertices of polyhedra need not only be thought in binary terms, as either present or absent. Different degrees of connection can be accommodated, and, for example, threshold values (or "slicing parameters") established, above which two points are seen as sufficiently connected (Gould 1981).

What emerges from all this is an "*operational definition* of structure" (Gould 1981). The simplicial complex made up of all the differentially connected simplices under consideration constitutes a "backcloth" forming the structure upon which "traffic" can flow (Gould 1981, 181). If the structural backcloth is a network of academics connected by their intellectual interests, traffic might take the form of ideas discussed, or book recommendations. If the backcloth is the global air travel system, traffic will obviously be flights. Thinking in terms of backcloth and traffic makes it possible not only to identify particularly well-connected nodes or hubs but also obstructions or "q-holes," that is, areas in the topological space devoid of connections (Gould 1981, 182). A final aspect of Q-analysis worth mentioning here is how it accommodates the phenomenon of change. Gould distinguishes between ordinary change in the exact nature or intensity of traffic along a structural backcloth that essentially remains the same, and more structural alterations of the geometry of connectivity itself. To stick with the example given earlier, the retirement of a colleague and her replacement by a different one with different interests would alter the polyhedral geometry of interests within an academic department. The closing of an airport would likewise constitute a structural change in air traffic.

Most of the remainder of Gould's 1981 paper reviews studies in geography and elsewhere intended to illustrate the widely varied ways in which Q-analysis can fruitfully be employed in research. Among the more interesting is the reinterpretation of a two-dimensional scatterplot of data points that would normally be subjected to least-squares regression. Gould effectively juxtaposes the multidimensional simplicial complex that describes the differential connections between the data points themselves with the simplistic series of disconnected one-to-one functional mappings that would be generated by a linear regression (Gould 1981, 184–185). The remaining illustrative cases need not detain us here. Gould closes the paper by returning to the larger claims that motivated it, asserting that Q-analysis "makes a mockery of the schizophrenic division between those areas of inquiry traditionally labeled Humanistic and Scientific" (Gould 1981, 196). In his view, "the backcloth-traffic distinction of q-analysis shows that human society, in all its aspects, is not embedded in a constant, naïvely simple geometry. . . . Rather it is embedded and defined by the changing geometries of multidimensional spaces" (Gould 1981). Q-analysis, Gould suggests, may thus be "the first example of emancipatory mathematics in the human sciences" (Gould 1981).

From Peter Gould to Alain Badiou

It is useful to compare Peter Gould's project of reconciling Heidegger and quantification with some aspects of Alain Badiou's project of grounding a radical and emancipatory philosophy in mathematics (Badiou 2006, 2008, 2009). Badiou, a former student of Althusser and Lacan, is unique among contemporary critical philosophers first of all for the depth and rigor of his mathematical training and competence. Even more striking, however, is the audacity of the connections he draws between advanced results of axiomatic set theory, number theory, and category theory, on the one hand, and on the other hand, important programs in the history of philosophy (Plato, Descartes, Leibniz, Kant, Hegel, Marx, Foucault, etc.) (Badiou 2006, 2009). More specifically, he asserts that axiomatic set theory constitutes a rigorous ontology of being-as-such, and that category theory provides a rigorous phenomenology, that is, an account of how beings appear together in concrete worlds. An overarching goal for Badiou is the vindication of a radical philosophy of the event that can make sense of such phenomena as political revolution, love, fundamental artistic innovation, or historic advances in thought.

Although the contexts, underlying philosophical perspectives and goals of Gould and Badiou are very different, there are also some striking parallels and similarities between the two projects. First, a brief sketch of some key differences. In historical terms, Gould was advocating Q-analysis as "emancipatory mathematics" at a time when critiques of quantitative spatial analysis were strongly oriented toward the dangers of technocracy understood in terms of state power. State power is also important for Badiou, but he takes more explicit account of the threats posed by what Shoshana Zuboff calls "surveillance capitalism" (Zuboff 2019). As one of his translators summarizes Badiou's position in this respect, "only if contemporary philosophy rigorously thinks through number can it hope to *cut through* the apparently dense and impenetrable capitalist fabric of numerical relations . . . without recourse to an anti-mathematical romanticism" (Mackay 2008, vii).

Not only the source of threat but also that which is threatened is different for Badiou than it was for Gould. In the geographic discourse of the late 1970s and early 1980s, humanistic accounts of the richness, complexity, and irreducibility of individual human lives are the main reference points for Gould's championing of Q-analysis. Although Heidegger's *Daseinanalyse* posed a challenge to rationalist, Cartesian and Kantian inflections of humanism, Gould's use of Heidegger is still essentially humanistic. Badiou, by contrast, is not in any conventional sense a humanist, and he is certainly not a Heideggerian. In fact, *Being and Event*, the title of the first volume of his *magnum opus*, as well as of the series as a whole, is a deliberate contrast to *Being and Time*, and to the fundamental claim that all knowledge is anchored in and relative to the lived temporalities and spatialities of human being-in-the-world. Whereas for Gould, it is a question of finding a mathematical language appropriate to the richness of human being-in-the-world, for Badiou, mathematical languages serve as the privileged means of attaining knowledge *not* dependent for its validity upon human beings.

Despite these fundamental differences, it is interesting to see how similar Badiou's mathematical phenomenology is to Gould's. I will first dwell on this similarity before briefly addressing some of Badiou's ontological claims that go beyond anything Gould contemplates. In broad terms, both Gould and Badiou take set theory as their starting point, although Gould does not freight set theory with the ontological significance Badiou gives it. They both then seek to describe the complex world of appearances in terms of relations between sets or elements of sets. It is in this second step that both of them implicitly or explicitly ground their claims that mathematics can be adequate to the infinite complexity and richness of the world. The category theory Badiou draws upon for this phenomenology, like the Q-analysis deployed by Gould, is fundamentally relational. The entities it works with are objects and relations or "arrows" between them (Badiou 2014, 13). In one sense, it can be understood as an abstract algebra of "mappings" that focuses upon structures of composition such as associativity or commutativity that obtain when mappings from some objects to other objects take place. Its level of abstraction allows category theory to be interpreted in terms of set theory, propositional logic, or topology (Badiou 2014, 183–216). The relational character of category theory means that anything or anyone described by it is portrayed "extrinsically," in terms of connections with other phenomena, not "intrinsically," in terms of inherent qualities or features (Badiou 2014, 13). Put differently, "what governs appearing is not the ontological composition of a particular being . . . but the relational evaluations that determine the situation and localize this being within it" (Badiou 2014, 167). In a broad sense, the same is true of the simplices and simplicial complexes described by Gould: they define individuals "extrinsically" through their relations with others. The frameworks used by Gould and Badiou are not equivalent: category theory has been worked out over a longer span of time by a large community of mathematicians, and has attained a more general formalization, so that, for example, attempts have been made to translate social network analysis (as noted previously, a close relative of Q-analysis) into the more rigorous notation of category theory (Dekker 2002). With this caveat in mind, it is still interesting to focus on the similarities.

Like Q-analysis, category theory enables an understanding of phenomena in terms of an abstract algebra, an ordering framework that makes possible evaluations of relations of "more," "less," or "equal," for example. For both Gould and Badiou, a central question is the degree to which two phenomena "share" something in the broadest sense. What is shared might be common interests, as in the example of members of an academic department from Gould discussed earlier, or, for Badiou, the "intensity" of relatedness of the appearance of different phenomena (Badiou 2009, 125). One of the first of many illustrative examples Badiou develops in *Logics of Worlds* is that of a rural French autumn landscape. A patch of red ivy glowing in the early evening sun is analyzed in terms of what its appearance has in common with the appearance of the wall on which it climbs, the roof above it, the distant hills and the sudden noise of a motorcycle skidding on gravel (Badiou 2009, 125–126).

The algebraic structure serving as the "domain for the evaluation of identities and differences in appearing" between the various phenomena in such a "world" Badiou calls a "transcendental" (Badiou 2014, 167). The qualitative meaning of "intensity" of appearance will vary depending on the nature of the world in question: to take a small sample from the wide array of illustrations Badiou discusses in *Logics of Worlds*, "intensity" may denote visual prominence within a painting, the military significance of contingents of fighters within a larger slave revolt, or the roles of groups involved in a public demonstration (Badiou 2009). In all cases, phenomenal intensity can be understood as degree of existence within a world, composed of degrees of coexistence and varying between a defined minimum value and an intensity capable of encompassing or "enveloping" all of the other phenomena with which something coappears.

A final similarity worth mentioning is the distinction both Gould and Badiou make between kinds of change. Both define routine, mundane, or everyday change, whether in the traffic moving along the backcloth of a simplicial complex in Q-analysis or political reforms within a "state of the situation" defined in a system of fixed subsets, as change that does not alter the encompassing fundamental order. Real change, by contrast, would take the form of a shift in the polyhedral structure forming the backcloth (Gould), or a shift in the state of the situation – phenomenologically, the transcendental ordering – brought about by the emergence of a new multiple formed out of present but unrepresented multiples (Badiou).

Conclusion: another concept of number?

In short, Gould's use of Q-analysis shares certain basic features with Badiou's philosophical vindication of mathematical representation: recourse to set theory, linkage of set theory with a relational algebra, and binary conceptions of change as either routine or structural. The crucial move made by both Gould and Badiou is to try to rescue mathematical abstraction as such from the negative associations it acquired in the 1960s and 1970s (Badiou's "anti-mathematical romanticism"

(Mackay 2008, vii)) by recourse to languages of relationality. For Gould it is only through engagement with abstract algebras that the variable dimensionality of individual human lives can be given adequate room. For Badiou, abstract algebras can reveal the order immanent to the coappearing of phenomena in a given world.

We might still ask, though, whether such relational approaches truly get us beyond the familiar critiques of quantification. Even if it is granted that Gould's polyhedral dynamics does not "crush" the dimensionality of human beings in the way linear regression or factor analysis do, and that Badiou's "transcendentals" do not illegitimately subordinate phenomena to a foreign standard of appearance, both still start with the reductive or "subtractive" registration and representation of complex human or nonhuman phenomena in quantitative (or quantifiable) terms. Indeed, precisely to the extent that what is subsequently done with the numbers is more subtle than traditional statistical manipulations, the *initial step of numerical representation* as such becomes all the more pivotal in deciding whether and how mathematics can have a place in critically aware social science.

This brings us to an utterly fundamental question, but one which surprisingly few critical scholars have asked. Gould does not focus upon this question, but Badiou does: "Isn't *another idea* of number necessary, in order for us to turn thought back against the despotism of number?" (Badiou 2008, 4). This question, perhaps even more than the specific answer Badiou gives to it, is tantalizing: Is the conception of number and numeration at the heart of familiar critiques of mathematics in social science fundamentally incomplete or misleading? Is it possible to conceive number in such a way that it does not "reduce," "crush," or reify what it purports to represent? Here I can offer only a brief indication of Badiou's affirmative response to these questions taken from his audacious attempt to provide a philosophical foundation for the concept of number in the book *Number and Numbers* (Badiou 2008). Starting with the idea that numbers, like all other beings, can be grasped via set theory, Badiou develops an ontological account of number that grounds all of the specific kinds of numbers discovered by mathematicians (including the integers, the rationals, the irrationals, and the reals) in a unified conception purportedly independent of epistemological assumptions.

The details of his argument are too involved to be summarized here. One key idea, however, deserves attention. Badiou argues that two numbers we think of as "greater than" and "less than" each other should not be understood as separate or distinct. In ontological terms, the numbers with which we count – or with which we are counted – (the ordinals) are instead related through *belonging*. A greater number in effect *encompasses* smaller numbers, in the sense that lesser numbers "belong to" the numbers that succeed them: "Order is always belonging" (Badiou 2008, 70, cf. 84). The concept of belonging, however, also carries associations of legitimacy in everyday parlance, and Badiou deliberately plays on these associations throughout his account of set-theoretic ontology (Badiou 2006). Being counted as belonging to a total can thus be understood as a form of "ontological" recognition of presence in a situation, of existence *prior to* "phenomenological" questions about specific attributes of individual beings or the officially recognized subsets in which something or someone is included (or not).

It is arguable that associations between being counted as present and such an ontologically inflected sense of belonging are at work, for example, in self-enumeration initiatives in the Global South, or in attempts by states to persuade members of national populations to be counted in censuses (Hannah 2021). By the same token, it is precisely the desire to limit the ontological recognition implicit in being counted as part of a population that has motivated conservative initiatives like the recent move by the Trump administration to add a citizenship question to the US Census in 2020. To put it schematically, neoconservative and right-wing populist logics tend to push for the production of ignorance about discrimination or disadvantage in relation to cultural or ethnic difference. Neoliberal logics tend to complement this push with the urge to render economic inequality invisible and, more fundamentally, to challenge the whole idea of a living human population as having any claim on resources simply by virtue of its "being present." To put it provocatively, we may be entering a phase in which being counted by states is no longer primarily a violation of the richness of human existence – although of course this possibility still exists – but rather increasingly one of the last "counterweights" to the ways we are counted by and for capital. Being-there for a state (even as a refugee or a homeless person) still often carries some inherent claim on the resources necessary to survive or even to live well. Being-there for capital does not.

To follow the line of argument developed here, perhaps two future roles for mathematically based human geography are, first, to use numbers to continue insisting on ontological recognition of the mere presence of human beings (and other living beings) in a given situation, and second, to experiment with algebras that can more sensitively depict the complex relationality of the human-geographic phenomena anchored in this basic presence.

Note

1 I am very grateful to John Pickles for his patient insistence that this paper must be the starting point for understanding Gould's larger project (Pickles n.d., personal communication).

References

Abler, R.; Adams, J.; Gould, P. (1971): *Spatial Organization: The Geographer's View of the World*. London: Prentice-Hall.

Badiou, A. (2006): *Being and Event*. London: Continuum, translated by O. Feltham.

Badiou, A. (2008): *Number and Numbers*. Cambridge: Polity Press, translated by R. Mackay.

Badiou, A. (2009): *Logics of Worlds*. London: Continuum, translated by A. Toscano.

Badiou, A. (2014): *Mathematics of the Transcendental*. London: Bloomsbury, translated by Bartlett, A. J.; Ling, A.

Barnes, T. (2001): Lives Lived and Lives Told: Biographies of Geography's Quantitative Revolution. In *Environment and Planning D: Society and Space*, 19(4), 409–429.

Barnes, T. (2004): Placing Ideas: Genius Loci, Heterotopia and Geography's Quantitative Revolution. In *Progress in Human Geography*, 28(5), 565–595.

Barnes, T.; Farish, M. (2006): Between Regions: Science, Militarism, and American Geography from World War to Cold War. In *Annals of the Association of American Geographers*, 96(4), 807–826.

Dekker, A. (2002): A Category-Theoretic Approach to Social Network Analysis. In *Electronic Notes in Theoretical Computer Science*, 61, 21–33.

Elden, S. (2006): *Speaking against Number: Heidegger, Language and the Politics of Calculation*. Edinburgh: Edinburgh University Press.

Gould, P. (1979): Geography 1957–77: The Augean Period. In *Annals of the Association of American Geographers*, 69, 139–151.

Gould, P. (1980): Q-Analysis, or a Language of Structure: An Introduction for Social Scientists, Geographers and Planners. In *International Journal of Man-Machine Studies*, 13(2), 169–199.

Gould, P. (1981): Letting the Data Speak for Themselves. In *Annals of the Association of American Geographers*, 71, 166–176.

Gould, P. (1985): *The Geographer at Work*. London: Routledge and Kegan Paul.

Gould, P. (1990): *Fire in the Rain: The Democratic Consequences of Chernobyl*. Baltimore: Johns Hopkins University Press.

Gould, P. (1993): *The Slow Plague: A Geography of the AIDS Pandemic*. Oxford: Wiley-Blackwell.

Gould, P. (2000): *Becoming a Geographer*. Syracuse: Syracuse University Press.

Gould, P.; Gatrell, A. (1980): A Structural Analysis of a Game: The Liverpool v. Manchester United Cup Final of 1977. In *Social Networks*, 2, 247–267.

Gould, P.; Johnson, J. (1978): *The Structure of Television Programming: Some Experiments in the Application of Q-Analysis*. University Park, PA: International Television Flows Project.

Gould, P.; Johnson, J. (1980): The Content and Structure of International Television Flows. In *Communication*, 5, 43–63.

Gould, P.; White, R. (2012): *Mental Maps*. London: Routledge, 2nd edition.

Hannah, M. (2021): Enumerating the Populace. In Withers, C.; Heffernan, M.; Domosh, M. (Eds.): *Sage Handbook of Historical Geography (2 Vols.)*. London: SAGE, volume 1, 337–354.

Hannah, M. (2001): Sampling and the Politics of Representation in US Census 2000. In *Environment and Planning D: Society and Space*, 19, 515–534.

Heidegger, M. (1977): Science and Reflection. In Heidegger, M. (Eds.): *The Question Concerning Technology and Other Essays*, trans. W. Lovitt. Garland Publishing, 155–182.

Johnston, R. J.; Harris, R.; Jones, K.; Manley, D.; Sabel, C.; Wang, W. W. (2014): Mutual Misunderstanding and Avoidance, Misrepresentations and Disciplinary Politics: Spatial Science and Quantitative Analysis in (United Kingdom) Geographical Curricula. In *Dialogues in Human Geography*, 4(1), 3–25.

Keylock, C.; Dorling, D. (2004): What Kind of Quantitative Methods for What Kind of Geography? In *Area*, 36(4), 358–366.

Mackay, R. (2008): Translator's Preface. In Badiou, A. (Ed.): *Number and Numbers*. Cambridge: Polity Press, i–xi.

Scott, J. (1987): *Social Network Analysis: A Handbook*. London: Sage Publications.

Smith, N. (1987): "Academic War Over the Field of Geography": The Elimination of Geography at Harvard, 1947–1951. In *Annals of the Association of American Geographers*, 77(2), 155–172.

Wyly, E. (2009): Strategic Positivism. In *The Professional Geographer*, 61(3), 310–322.

Wyly, E. (2014): The New Quantitative Revolution. In *Dialogues in Human Geography*, 4(1), 26–38.

Zuboff, S. (2019): *The Age of Surveillance Capitalism: The Fight for a Human Future at the New Frontier of Power*. London: Profile Books.

14 Conclusion

A virtual discussion about the quantitative revolution's legacy for past, present, and future

Trevor J. Barnes, Luke R. Bergmann, Ferenc Gyuris, Matthew Hannah, Mariana Lamego, Larissa Alves de Lira, Michiel van Meeteren, Boris Michel, Olivier Orain, Katharina Paulus, Matteo Proto, and Guilherme Ribeiro

Instead of a conclusion by the editors and a futile attempt to establish one common denominator of these histories or to just repeat the diversity, we decided to close this volume with a virtual discussion. This virtual discussion was written during a time when actual in-person discussions were difficult due to COVID-19 and when the notion of an in-person meeting, such as the workshop that gave rise to this book project, seemed impossible.

What is your favorite historical quote about the "quantitative revolution" from your geographical tradition? What does it mean to you?

Mariana Lamego

> *Undoubtedly, quantitative methods should be employed as much as possible so that the results can be measured and compared. Each day, conclusions based exclusively on field observations and analysis of examples become more contested due to the high degree of subjectivism and the impossibility of measuring the degree of generalization of the examples taken. It is not enough to describe a fact or to prove its occurrence in some area. It must be quantified, delimited according to precise criteria and perfectly characterized. Only then can more accurate comparisons be made with other similar facts and be classifiable in different degrees of proximity.*

This quote is obtained from a paper written by two professors from Rio Claro University, Antonio Oliveira Ceron and José Alexandre Felizola Diniz (1971). The paper is a pioneering text in the introduction to and application of quantitative techniques in geographic research. Published in 1971 in the Revista Brasileira de Geografia, the paper is the primary vehicle for disseminating geographical research by the Brazilian Institute of Geography and Statistics. The text clearly expresses the enthusiasm for the arrival of the quantitative revolution's echoes among Brazilian geographers. Unlike a more spatial spread of the UK and USA quantitative revolution, in Brazil, the incorporation of

DOI: 10.4324/9781003122104-14

quantitative novelty was very local and concentrated in two distinctive venues, which helps us understand the prominence of two versions of Brazilian quantitative geography, as I pointed out in my chapter.

Matteo Proto

I would quote Lucio Gambi's (1971) words reported at the end of my chapter:

> Not even so-called quantitative geography led to better ideas; despite its understanding as a new dynamic aimed at introducing a traditional research approach for contemporary problems, it could not hide its intimate nature, i.e., processes that aim to transform human beings (destroying their historical value) into androids and human works into artifacts, in keeping with the operating plans of a delimited technocratic class.

Beyond its critique of quantitative geography's reductionism, this quotation suggests the significant bond between the development of these research methodologies and the political and economic context of postwar Italy (and Europe).

Michiel van Meeteren

I'm going to go for the quote that grasped me when working on the chapter for this book:

> *Although we can congratulate the writers with the fact that they dared to introduce a new research method, it is to be regretted, that they have silently ignored everything that has already been previously done in this area. Too much iconoclasm, too little reform.*

The quote, translated from Dutch, is from a book review by Willem Steigenga (1950) about a spatial-economic survey on the Dutch city of Amersfoort (Klaassen et al. 1949). The survey was innovative in its use of econometric methods. The study's lead author (Leo Klaassen) would become an important early cheerleader of Walter Isard's regional science project.

The quote signals important themes in my own project on reworking geography's history (Van Meeteren and Sidaway 2020). First, our historiography has in the past too often emphasized radical change over continuity, "silently ignoring earlier work." Resultantly, we often lost track of the interesting precedents that allow us to track evolution in the discipline. The irony is, of course, that it is now my job to help rescue Klaassen et al.'s study from oblivion. Second, I think we need to take Steigenga's remark on iconoclasm to heart. The focus on and glorification of radical breaks and turns, "iconoclastic moves," in disciplinary development makes it harder to reform the discipline. If you first trash your predecessors' work, how can you expect them to adopt some of the ideas you try to convince them of? Instead you get a fractured antagonistic field, which is detrimental to disciplinary development.

Ferenc Gyuris

> *Over the course of the last decade, it has become apparent that [the choice between] quantitative or empirical geography presents a false dichotomy. The application of state-of-the-art mathematical-statistical methods has produced barren results and remains drowned in empty formalism when those who apply it are not even familiar in a qualitative sense with the subject on which they have collected and processed data.*
>
> (Enyedi 1984)

This is how György Enyedi (1930–2012), the author of several pioneer quantitative studies in Hungarian geography during the 1960s, summarized his view two decades later in 1984. I like this quote for its twofold meaning. It aptly highlights the deep cleavage between the "hype" initially surrounding quantitative geography in many countries and the questionable geographical relevance of many studies it actually ended up in. However, the quote also suggests, even if less directly, that a balanced combination of quantitative and qualitative methods, data and knowledge can produce fruitful results. This means to me that quantitative geography is a false alternative, but it can be a relevant and useful extension of "nonquantitative geography."

Trevor J. Barnes

> *[Geography's] quantitative revolution has run its course and diminishing marginal returns are apparently setting in as . . . [it] serve[s] to tell us less and less about anything of great relevance. . . . There is a clear disparity between the sophisticated theoretical and methodological framework which we are using and our ability to say anything really meaningful about events as they unfold around us. . . . In short, our paradigm is not coping well. It is ripe for overthrow.*

My favorite quote about the quantitative revolution comes from David Harvey (1972), which is not a celebration of it but is one of the first nails in its coffin. I first read this quote in 1976 as a second-year undergraduate at University College London. I was doing a joint degree in geography and economics. I already had done a couple of courses in statistics and was being exposed both to the abstract theories of economics and spatial science. I thought it was all very arid and unreal as well as dull as dishwater. It was not what I came to university to do. And then I read Harvey and this quote. It was an epiphany. I wanted to sign up on the dotted line right away and to be a geographer like him: to overthrow the paradigm and do something relevant. It didn't quite work out as I imagined, but it is a quote that made me want to continue with geography and which I continue to find inspirational.

Guilherme Ribeiro

> *From a Positivist point of view, answers and solutions can be wrong and modified in the next; one seeks to improve them and to verify their validity by refutation. From a Marxist point of view, propositions can be neither verified nor refuted. They are dogmatic, and reply and solution are more important than the problem. Marxist solutions to the problems of the world are already ready.*
>
> (Christofoletti 1982)

The premise of this quote means that we cannot transform our theoretical basis in a straitjacket. All of them are uncompleted, and science is an activity that is inherently political. From textual elements to the final results, no scientist is neutral. However, many times we became blind to our methodological preferences. As a result, only the other methods are dogmatic. In his scientific defense, the geomorphologist Antonio Christofoletti reacted against a political and ideological emphasis proposed by Marxism in the name of a stretching science that he found in Positivism.

Matthew Hannah

My response has its origins in the dominant but conflicting influence upon my perspective exercised by the late Peter Gould, my doctoral advisor, and the main subject of my contribution to this volume. As noted in that chapter, Gould was one of the most provocative and combative champions of the Anglophone "quantitative revolution," and therefore a rich source of potential "favorite quotes." Not surprisingly, most of the quotes that occur to me relating to the quantitative revolution stem from Gould. However, Gould's love of provocation and verbal combat led him on many occasions to cross the line – at least at the rhetorical level – into reactionary politics and the abandonment of all respect for his opponents. Thus many of his most quotable quotes either dismissed whole traditions or perspectives out-of-hand (for example, his writings on the "Augean Stables" or later, his screed against "cathartic geographies") or lionized the "heroes" of the Anglophone quantitative revolution as intrepid "men of science" (see The Geographer at Work).

Instead of quoting from his demonizations and hagiographies, I want to quote Gould from the classroom. For decades, he taught his introduction to quantitative methods at Penn State on Monday, Wednesday, and Friday mornings at 8 am. He deliberately chose this timeslot in order to impress upon the students the necessity of effort in learning quantitative methods. To emphasize the need to buckle down and think hard at such an early hour, he repeatedly urged his students to "wrap a cold towel around your head!" Having taught geography myself now for more than two decades, I belong to a generation that has witnessed a serious weakening of the role of quantitative methods in human geography curricula. Now it is common for human geography students to avoid spatial analysis like the plague if they possibly can. This is a great loss, especially in an age when issues such as segregation, gerrymandering of voting districts, and spatial inequalities of all kinds are on the agenda in many countries. Human geography needs to retrieve Gould's "cold towel" from the laundry hamper and put it to work again.

Larissa Alves de Lira

The quote I would like to highlight is that of Brazilian geographer Speridião Faissol, a member of IBGE and one of the stimulators of the employment of quantitative geography in Brazil. He makes the following observation in 1960 at the 20th General Assembly of the National Council of Geography:

> *the present conjuncture of the country offers the inspirations and the opportunity for modern geographic science to place itself effectively, as it should, at the service of the administration, in the continuous and permanent objective of promoting the common good, for the service of the administration of this 'contemporary of the future', as Congressman Pinheiro Chagas qualified to President Juscelino Kubitschek de Oliveira, who decisively established the empire of numbers, analysis and regional planning in the government of public affairs.*
>
> (Faissol 1960)

President Juscelino Kubitschek promised some important social reforms (of progressive bias) in this country, which was divided between the people and the oligarchy. From this quote, a person can infer how quantitative geography in Brazil could be associated with progressive aspirations for that time. Furthermore, I suggest that quantitative geography "arrived" in Brazil during a movement that began with external influences. Thus, its arrival intrigues me regarding the question of its adaptation to this territory. I link quantitative geography to several movements, among which we can highlight: the degree of humanization of the territory and the advance of the economic frontier, and on the other hand, the technical investment of the community of geographers, such as the formation of solid institutions, teams of geographers prepared to go to the field, or the development and placement of computers at the disposal of geographers. None of these movements are homogeneously spread throughout Brazilian territory. Therefore, I think that some geographers who proposed a quantitative geography in Brazil carried some dose of *utopia*, either for a fairer country in some cases, or the idea that this geography was a geographical paradigm that could be easily adapted to the geographical community. In this sense, the present was a "contemporary of the future." To what extent it remained more "future" than "contemporary" in the history of geography in Brazil is a topic that puzzles me.

Boris Michel

The first paragraph of Walter Christaller's (1933) "Central Places in Southern Germany" presents some of the lesser discussed peculiarities of early quantitative thinking.

> *In inorganic and organic nature, the crystallization of mass around a nucleus is an elementary form of the order of things, which belong together- a centralistic arrangement. This arrangement is a mode of thinking that exists not only in the human world of imagination that emerged because of man's demand to order but also out of the inherent laws of matter. The same centralistic principle is also found in certain*

> *forms of human community life and certain sociological formations and is materialized in a visible objectification of said way of life.*
>
> (Christaller 1933, own translation, the 1966 English translation has some inaccuracies and loses some of the connotations if a person finds it interesting. Additionally, the German term for visible is translated as "invisible")

While just a starting point for his elaboration on centrality and some foundations of urban geography, this paragraph brings together a number of topics that I find interesting. These topics refer to physical laws as the basis for social laws and uses – as many quantitative geographers did – the authority of physics to make claims for the existence of these social laws. The language also resembles the visual epistemology of classical geography. It is true because it can be seen. I think this mode of using a visual language is a kind of bridge between classical regional geography and abstract mathematization and quantification. We see this often, for example, in Bunge's "Theoretical Geography."

Katharina Paulus

I choose a quote by the German geographer Eugen Wirth. The author, whose research can be considered classical regional geography but who has also been criticized for using his spin on the quantitative turn of geography as a "conservative embrace of the "Revolution" (Bartels 1980). With his book "Theoretical Geography," Wirth was indeed trying to put his own stamp on what is happening within the discipline by using the appropriate vocabulary and developing a take on quantitative and systems theorists – mainly to keep regional geography alive in a new guise. For the last sentences of his book, however, he abandons the efforts of his embrace to remind his fellow geographers of what actually constitutes geography: "Despite all the enthusiasm for theory, mathematics, computer simulation, abstraction and logical calculation, one should never forget that for a geographer, in addition to scientific work, a sunrise in the high mountains or a day in the tropical rainforest, the still untouched building stock of a historical old town or the scenery of a classical Mediterranean coast, the natural spectacle of an active volcano or the pulsating, effervescent life of a cosmopolitan city such as Istanbul or Rio de Janeiro can be among the fascinating experiences for the sake of which it is worthwhile to be a geographer" (Wirth 1979).

Olivier Orain

I do not have a favorite quote about the "quantitative revolution." Gathering or handling quotes appears to me to be something you do when you seek inspiration. Quotes show that you are an insider. Quotes hold a performative, unifying value; they can be used to bring together an entire collective. In this respect, they are similar to the properties that Thomas Kuhn reads in "shared values" performed via aphorisms. My historical research focuses on scientific fields in which I am not involved. I have never been a "quantitative geographer," even though I taught spatial analysis for many years at Université Toulouse Le Mirail (now Toulouse Jean-Jaurès), as mandatory knowledge every aspiring geographer should have. So, I never looked for quotes that I could endorse.

However, to a historian of science such as me, *emic* quotes are worthy of attention (and work on) when you seek inside expressions of various symbolic generalizations that help build a scientific community. I recall Denise Pumain using Brian Berry's phrase "city as system within a system of cities" (in English) as a motto for her research and teaching in the field she dubbed "theoretical urban geography." This application is symptomatic of a cultural appropriation: in Berry's famous (only in France?) 1964 paper, there is no systematic use of such a phrase. However, in Denise Pumain's work, her use of this phrase was considered high value to characterize her approach.

From an *etic* perspective, I use quotes when I want to prove something or when I analyze the language that one or many geographers use. I do not believe that they have emotional significance; they are an episodic material to show or to dissect.

Has your understanding changed since you started thinking about local and historical varieties?

Ferenc Gyuris

The history of geography is full of exciting events. Nevertheless, the "quantitative turn" seemed to be something special, even mystical to me. I thought it constituted a historically unexampled radical conceptual and methodological shift in our discipline. This was a result of two influences. The first influence was the narrative of some regional scientists who I became familiar with during my university studies and who positioned regional science along with its quantitative and model-centric approach as a "truly" scientific, and thus superior, discipline, in contrast to geography as just a "descriptive" and not truly scientific field, rather a "general education subject." Another influence was that of mainstream Anglophone narratives – not only those the US "pioneers" of quantitative geography created about themselves but especially those in post-positivist critical works, which interestingly, reinforced this image even if in a critical tone. Today, I see things differently. The "quantitative turn" is one of the many remarkable changes in the discipline's history; it is an exciting one and is truly significant; still, it is just one out of many, which have had a very uneven impact on different institutions and individual careers. My understanding of what the terms "quantitative" and "nonquantitative geographers" mean also changed, for I realized that many scholars who have never truly identified themselves as "quantifiers" and who applied diverse approaches and methods had a very important role in fruitfully introducing a quantitative mindset into geography.

Trevor J. Barnes

My view changed from the late 1990s when I first started to collect oral histories of the quantitative revolution from "pioneers" at different sites within North America – Seattle, Iowa City, Chicago, Evanston (IL), Columbus (OH), and Toronto. Before I started that work, I believed that the quantitative revolution was undifferentiated. Science was universal, I believed, so the quantitative revolution was universal too – the same everywhere. I imagined the quantitative revolution as like a wave rolling over space, leaving in its wake uniformity, everyone doing geography the same way. But from doing my oral history interviews with people at various North American geographical centers of calculation that clearly was not true. The experiences of those who participated and the contextual circumstances in which it occurred were vastly different from place to place. Place mattered; geography mattered. I should have known that already. What geography brings is contingency and variability. It makes place always a black spot when trying to uphold supposed universals.

Guilherme Ribeiro

It may be useful to recognize the different global impacts of the "quantitative revolution." French regional geography is much more rooted in Brazilian tradition than spatial science; the same is not valid in the United States, for example. As there is no science out of place, it is interesting to observe the circulation of ideas, the main journals and institutions from many countries, the reach obtained by a book or a paper, and mostly, the different materials translated to disseminate the novelties. Local and historical varieties allow us to draw a more complete picture involving scholars, concepts, and approaches *in common*, which is crucial to shaping a transnational history of geography.

Mariana Lamego

A place-based narrative on the arrival of the quantitative revolution among Brazilian geographers reveals the crucial role of an international network of human and nonhuman actors responsible for disseminating new theories, methods, and techniques in geographical investigations. A positioned and embodied history of quantitative geography in Brazil made it possible to overcome canonical, essentialized, and superficial narratives on Brazil's discipline development.

Michiel van Meeteren

My understanding of geography's quantitative revolution changed the moment I stopped relying on what textbooks said, often in very critical terms, about the quantitative revolution and started reading the original papers and books instead. Many geography textbooks have become so accustomed to describing this period of geography as a "cautionary tale" or "cul-de-sac" that before you know it, you only understand a caricature.

This was exacerbated by my local background. In the Netherlands, there has not been a big disruption between, for instance, economic geography and regional science. Quantitative methods just had their place in the "method toolbox." But as the discipline became internationalized in the Netherlands (i.e., it became English-language dominated), we replaced our Dutch language textbooks with English language ones and all of a sudden quantitative geography was problematic. The incongruence between the anglophone textbook and local Dutch practice definitely drove some of my research in this field.

Matteo Proto

Actually, before starting to inquire about the origins of quantitative revolution in Italian geography, I thought that the development of this paradigm was just related to the advancement of spatial sciences and the development of computing devices in North America and Northwestern Europe. I did not consider the connection with the quantitative methodologies developed in Italian and European geographies since the late 19th century and especially the bound with research in economic geography. The second aspect I did not consider regards the rise of a critical reflection in geography in the late 1960s/1970s and its connection with quantitative geography. Similar to international figures, such as David Harvey or Gunnar Olsson in Italy, the introduction and implementation of quantitative methodologies and theoretical frameworks pushed for a critical reflection and the deconstruction of the models embedded in geographical thought. Geographers could then move toward critical thinking and reject any positivist theoretical scheme due to the awareness of the gap between the quantitative theoretical and methodological frameworks and their application to the research.

Matthew Hannah

Yes, certainly. In earlier retrospectives on the quantitative revolution an uneven geography of knowledge production was already thematized alongside the more "diffusionist" narrative of the "spread" of spatial analysis from Anglophone lands to "others" via publications or guest professorships. Sweden, and particularly the Lund School around Torsten Hägerstrand, figured prominently as sources of distinctive traditions that qualified this model. However, I (and undoubtedly many others) for a long time simply thought of the diversity of traditions as being absorbed into a transnational, cosmopolitan discourse itself still largely carried on in the Global North.

Learning about different trajectories of quantitative geography in Brazil and elsewhere has been one important source of new insight complicating the cosmopolitan picture for me. The ground for the recognition of this diversity was already prepared, in a sense, by the work of Trevor Barnes and his collaborators in "provincializing" the Anglophone quantitative revolution, in showing how specifically it was shaped by the Cold War and other geohistorical shifts.

Larissa Alves de Lira

I believe that all the stories of geography should be considered "situated knowledge" and that the history of qualitative geography has much to gain in its insertion in this movement of "spatial turn." At the same time, this is a movement that is closer to the construction of the global in the history of the discipline. Winning in variety, it is as if we gained many histories for one history.

Katharina Paulus

Before this project, we assumed that all the local geographies basically translated the now hegemonic geography into their contexts and that traveling theories meant implementing those into complex local contexts. This translation could mean a selective reading and application of certain ideas suitable for their context and the disregard for other ideas. For German geographers, for example, one challenge was to establish a clear demarcation from the statist blood-and-soil geography of the Nazi era and its take on spatial planning and simultaneously to be perceived as experts for spatial planning. The translation could also mean to find ways to navigate institutional power structures by giving in to some of the old geographers and geographies. German geographers, for example, often tried to keep intact some of the wording and content of classical regional geography (Länderkunde) and were hardly able to propose a more radical break (the previous quote by Wirth is an example). However, what became increasingly obvious was that all these local quantitative geographies were also building local quantitative geography histories. While most surely acknowledged the role of Anglophone geography and Anglophone geographers, most local histories were also able to knit tight connections to local precursors.

Olivier Orain

First, I understand the question as something you ask a practitioner when he stops to practice and starts thinking about his practice, exploring the past (or diversity) of an activity and trying to decipher a "philosophy" from it. In this respect, I cannot answer the question, as "thinking about local(ly) various and historical activities" was my perspective from the beginning. The geography I learned as a student at the university was quite traditional (and very boring). I "encountered" spatial analysis first in 1991–1992 during the last year of my master's program, but it remained vague at the time. It is only through my experience teaching in Toulouse (from 1996) that I educated myself on these topics, already from a historical and epistemological perspective, even though Denise Pumain's writings have served as a major liminal frame.

Therefore, "quantitative revolution" appears to me to be another emic expression (see also the uses of "quantitative" by people who are critical of new methodological trends in research, but in this case it was not associated with "revolution" but sometimes with the "so-called revolution"). If an event similar to a revolution has occurred, in a Kuhnian's (etic) understanding, it is in the way geographers thought and constructed their objects since the mid-1950s. Geography was already "quantitative" in some ways, using statistics, elementary calculus, etc. What changed, in my opinion, is that "spatialists" started, as pointed out by Edward Ullman (1954), to focus on interactions in a constructed system of spatially distanced places, instead of trying to explain why a "site" is a particular location and not anywhere else. The birth and rise of the concept of "space" to the core of geographic theory is what constituted a "gestalt shift" in "quantitative geography," a phrase that was probably more socially acceptable and attractive at the time. However, Bunge's PhD (1962) was about "theoretical" – and not "quantitative" – "geography."

How would you describe the legacy of this history, internationally and nationally?

Ferenc Gyuris

In my understanding, the history of the "quantitative revolution" refers to both a "spatial science" approach and a mathematical-statistical toolkit. Although both went hand in hand from the 1950s onward, I do not think that they necessarily have to be lumped together. As far as I can see, quantitative geography was pushed to the periphery of the discipline in those local and national traditions where postpositivist approaches emerged as early as the 1970s or 1980s. This peripherization, however, concerned not only the formalized model-centric approach but also very often the use of "numbers" (quantitative data) and even the analysis of macroscale issues. At English and German-language geography conferences, many times I felt as if geography should be microfocused and purely qualitative, or it would otherwise inevitably run into the trap of being "uncritical" and "exclusionary." I think it is not so. Instead, if geographers refrain from painstakingly discussing such topics, it will just open the door for "armchair geographers" to bring back and popularize utterly problematic views from the far past of the discipline, e.g., naïve environmental determinism, which is easy to find in contemporary bestsellers on geopolitics. On the other hand, where quantitative approaches have maintained a strong position until now, studies still often suffer from massive empiricism and the lack of social theories and social contextualization. In my view, it would be important to find a more balanced in-between approach, so that neither the baby gets thrown out with the bathwater nor the bathwater remains in the tub.

Luke R. Bergmann

Much remains to be done to trace the ways in which the milieus of *Theoretical Geography* and of other early quantitative geographies drew upon, and have been interpreted by, scholars working in many languages and in many parts of the world. Recently, the fraction of citations to *Theoretical Geography* from pieces written in English is little more than half; this observation is suggestive of ongoing conversations and should be seen to provincialize my efforts here. In our chapter, we do make various references to ways in which recent decades of anglophone geography have spoken of *Theoretical Geography* and the milieu in which it was written by Bunge. One use of this history has been in a recounting of disciplinary developments that necessarily led to the quantitative geography and GIS of the present. Yet we write of how our own readings allow *Theoretical Geography* et al. a more diverse set of interpretations. It is perhaps not by accident that we examine the complicated relationship between Bunge and Hartshorne, a relationship that was not solely external and opposed (as the Schaefer–Hartshorne debate has been understood) but as an internal relation of generative contradictions that were important both to Bunge and to what Bunge's geography has offered others. It remains an undisciplined geography, a geography that is neither reducible to nor separable from mathematical thinking, a geography engaging life while committed to theory, a geography whose spaces are quantifiable yet multiple and produced, and a geography whose science is always-already political. It is a geography whose legacy offers challenges to our histories of geography, a geography born of crises, a geography of hope.

Mariana Lamego

I believe that the legacy of the history of quantitative geography in Brazil was the openness of new international partnerships and the consolidation of intellectual exchanges with mainly Anglo-American geography. Brazilian geography has its intellectual roots deeply connected with French geography since the institutionalization of the discipline. The quantitative revolution enabled the enlargement of the disciplinary map and circulation of researchers. Many scholars from the USA and UK came to teach at Brazilian institutions. On the other hand, the quantitative revolution opened new paths for students and scholars from Brazil, who visited departments in the USA and UK to learn and share a new set of techniques and practices.

Larissa Alves de Lira

Internationally, the legacy of quantitative geography, at any of the moments of its tradition, will always be significant because it has seriously proposed the debate of quantification throughout human geography. As far as Brazil is concerned, I think it is also interesting to consider this tradition in the interior of a relation of center and periphery. Could mathematical and technical models of geography originate and expand from a country on the periphery of the world economy?

Guilherme Ribeiro

This kind of intellectual exercise is difficult due to this strong degree of generalization, but I presume that the current generation was seduced by a unilateral explanation about "quantitative revolution." From a Brazilian perspective, quantitative followers were relevant to maintain an established space within the federal sphere represented by the Brazilian Institute of Geography and Statistics (IBGE in Portuguese), for instance. Nonetheless, as their golden age is directly linked to the military dictatorship, it is very difficult to rebuild a consecrated image marked by geographers without social conscience in a peripheral country such as Brazil. This question can only be answered when we understand that the history of geography is neither an apolitical field nor a far west movie polarized by villains and heroes. In contrast, shall we cancel all geographers with whose political inclinations we disagree?

Matthew Hannah

I interpret "this history" not merely as the events and processes described in this volume ("history as it happened") but also and inextricably as the way they have been understood ("history" as the way we interpret what happened).

One important and damaging legacy of the conventional, cosmopolitan, or universalistic view of the quantitative revolution, whether this view was held by its champions or by its critics, has been to identify quantitative methods too exclusively with "science" and "rigor." This has made it more difficult for proponents of quantitative methods to recognize and appreciate the equally valid forms of rigor exemplified in many qualitative methods, in particular in the form of discourse analysis, systematic reflection, and analysis of positionality. At the same time, critics of quantitative methods have often been unable to recognize continuities and connections between qualitative and quantitative forms of rigor. As a result they have failed to appreciate the continued viability and relevance of statistical methods within a critical social-scientific research program. The new, more "genealogical" understanding of the quantitative revolution presented in this book emphasizes the contextual embeddedness of different developments of quantitative methods in geography, and thus should make bridging this gap easier.

Matteo Proto

Well, this is quite a difficult question, which deserves more than two paragraphs. I think we could outline two different perspectives. First, related to the previous answer, the legacy of quantitative revolution could be considered in its role of advocating critical geography. In this sense, the critique of quantitative geography and calculative methods developed by human geographers progressively increased the gap between hard sciences and human/social sciences. This situation is still problematic despite the recent development of more than human and vitalist approaches that aim to merge the human and nonhuman dimensions and to embrace life sciences methodologies in the geographical discourse.

On the other hand, the development of quantitative methodologies in geography went well beyond the calculative methodologies of the 1960s and 1970s, especially with the progress in the field of digital cartography and GIS since the 1980s. This is true both at the international level and in Italy. I just quote as examples the theories (or lack of theories) of the so-called new geography/neogeography that are also reflected in Italy (even if much more in the field of urban studies than human geography) and the development of landscape ecology, as well as the systemic investigation of the landscape based on biological and geomorphological theories and methods, but also integrating human aspects.

Katharina Paulus

In my opinion, no clear picture emerges in regard to the long-term effects. While German geography likes to tell the story of quantitative geography as one of the truly revolutionary events that shook the discipline in 1968 and 1969, the 1970s and early 1980s are often described less as a period of postrevolutionary excitement than as one of restoration and monotony. This depiction may be due to the structure and dynamics of German universities, where many professorships were newly filled during the expansion and upheaval of the 1970s, which then focused on aspects that were relevant in educating "Diplom" geographers. The 1970s and 1980s were also characterized by a widespread blockade of radical and critical thinking in geography. Compared to other disciplines, geography remained relatively conservative. Therefore, it is not surprising that critical geography, after it emerged as a recognizable movement in the early 2000s, explicitly positioned itself against quantification.

Michiel van Meeteren

Objectively, the legacy of the local version of the quantitative "revolution" in Dutch and Belgian geography, the two contexts where I was educated in, is very limited. It simmers in the common sense of locally trained geographers, although as the generation of people that studied in the 1970s and even the 1980s retires, this common sense is eroding fast. The one exception may be in evolutionary economic geography, which builds on a strong Dutch lineage through the oeuvre of Jan Lambooy rooted in the Dutch quantitative turn. However, I do wonder to what extent people realize how far these ideas go back.

For me, it makes it all the more fascinating to study these legacies. My PhD (Van Meeteren 2016) is literally using the philosophical ideas of the Belgian "new orientation" (Saey 1968), on contemporary Belgian settlement geographies. Just to see if the old ideas are still useful. And they were remarkably applicable to understand aspects of today's world. So the challenge is one of curation. Study this legacy. Make it accessible. Create roadmaps so that people know where the good stuff is buried.

Olivier Orain

I guess you mean the legacy of the first decades of spatial analysis? This is a question I would like to answer if I actually knew what it is that "spatialists" do nowadays (but I do not). However, as a historian, I do not like the word "legacy," as it shapes the past as a virtual treasure for the present. I can provide a brief answer about the knowledge of French "historical" theoretical quantitative geography (TQG) among current students and young researchers: I fear that the specialized students in the master's program where I teach have minimal knowledge about this topic. The history of geography is scarcely taught in France, and "spatial analysis" is taught in some geography departments as a set of "techniques" useful for GIS or research, not as a historical development. Thus, what remains from the past is a few names and frames, such as "Von Thunen's model," "Gibrat's law," or the famous Haggettian drawing (1964) of the five (or six) stages in locational analysis (which are actually a theory of the organization of human space).

Students and young researchers do not know any French "classics"; they have not read them and are not familiar with Brunet's "chorèmes," even if they vaguely know his name. This situation is just one example. Students who are trained in "spatial analysis" curricula have slightly more knowledge about the past or these "famous" names. However, to ascertain this information, I would need to carry out a survey among them. My courses are where my own students learn about such a history, but I do not have enough time to teach it. My colleagues at Université Paris 1 Panthéon-Sorbonne might shed some light on a specific model or theory but from an emic perspective.

Currently, "quantitative" techniques comprise highly specialized knowledge that you can barely master in its whole. Many productions of the past have been reincorporated in actualized discourses, formalizations, and images. However, a dominant "presentist" spirit transforms them into a set of modern interpretation systems emptied from their historical relevance. Again, I would say that such a hypothesis should be tested with a rigorous survey.

How would you describe the relevance of this history's legacy for new quantitative methodologies, for example for discussions about digital geographies or big data?

Matteo Proto

As mentioned in the previous answer, the development of new quantitative methodologies did not interrupt in the last three decades, despite the decline of quantitative geography as a main discourse. Recent accounts of the progression of quantitative methods (Johnston et al. 2019) report a substantial linear progression of theoretical and methodological frameworks from the 1960s to the present, despite reclaimed but uncleared philosophical shifts. Regarding critical studies, proposals of critical GIS do not seem to have produced any significant useful tools. Similarly, the postrepresentational reflection on cartography (e.g., Kitchin et al. 2013) actually produced an evolved and advanced critique of cartographical reason and mapping rather than suggesting new cartographic methodologies. Therefore, I would synthetize the legacy of quantitative geography by suggesting, the substantial progression and lack of criticism toward this legacy in the contemporary discussion developed by scholars that currently employ quantitative methods. On the other hand, critical scholars find it difficult to employ quantitative methods and data. This situation is also evident in the Italian context, where quantitative methods and representational tools are mainly employed by engineers in the field of geomatics but often neglected by human geographers.

Ferenc Gyuris

I like eclecticism and "Anything goes!" from Paul Feyerabend (1975). I think the diverse heritage of quantitative geographies confirms that every approach, method, and type of data also has its relevance and limits – not "either-or." In 2021, I regard it as "given" that human society produces a shocking amount of quantitative data every day, and these data can be analyzed from a geographical perspective. If we abstain from such investigations because we are afraid of obtaining sensitive results that require very careful contextualization, these data will still exist, with a general social interest in them. Someone else will perform the job, and maybe in a way that is much more problematic than we could imagine. In addition, I consider digital geography and big data exciting methods, similar to maps, which can be combined with many approaches and used in different ways. Thus, I think that these new methods and the surrounding debates provide a great opportunity for highlighting social issues and their underlying reasons more understandably than what would be possible via more conventional means.

Boris Michel

I often feel the (somewhat pretentious) urge to argue against current "digital geographers" and current "digital geography," which is not new; much of this field was developed in the 1960s. For example, in the new interest in social physics and datafication, an interest that is often disregarded is the notion that there have been at least two social physics in the last 200 years. In addition, something similar could be said about the current feeling of a data deluge, which is also an old issue for historians of science. However, what I find increasingly interesting and relevant is the extent to which concepts we often think of as historical are very recent. My favorite of these concepts is probably "data." It is relatively recent that geographers think about the things they encounter as data and I am still not sure what geography before data looked like.

Guilherme Ribeiro

From Portolan charts to Google Earth, mapping is a rational procedure that was revolutionized by digital technologies at the beginning of the 21st century (Brotton 2012). In a world mastered by neoliberal capitalism, technical tools have had a crucial role in the process of increasing value to the space. In this sense, quantitative geography was not an exception but a particular chapter in the long-run history of space and time. Currently, when classic references of meters and hours have been displaced by nanomeasurements and micromeasurements (Virilio 1984), "digital geographies" as "big data" emerged as the new faces of this history. Is there any similarity between these novelties and the quantitative revolution? Perhaps the risk is transforming existences, hunger, migrations, and socioeconomic inequalities in vectors, points, averages, clouds, and statistics in the name of consumption, stock markets, political manipulation campaigns, urban and regional planning from above, and so on (Santos 1978). On the other hand, we cannot overlook that geographical science is able to explore new technologies.

Larissa Alves de Lira

The use of numbers has led to many controversies in the history of geography. Which of the geographical traditions has not given rise to controversies? However, as many quantitative geographers pointed out, there is no science without quantification. What geographers do with quantification and the theories they elaborate is not only a scientific problem but also a political and social context. Previous traditions of qualitative geography have been important for today's traditions, not only for the invention of their techniques but also for the whole philosophical, political, and ethical debate that accompanied them.

Mariana Lamego

Technical and historiographical issues necessarily limit the legacy of Brazilian quantitative geography. Since the beginning of applying quantitative techniques in the country, the difficulty of obtaining modern computers has hampered modern regionalization and planning techniques. On the other hand, the historical version that associated these approaches with the military dictatorship and the state's arbitrary action also made it difficult to deepen more consistent technical legacies. Despite these factors, it is interesting to observe how quantitative techniques stimulated a pioneering geographic information system application.

From the encouragement of the introduction of quantitative techniques, the first computers were used to perform spatial analysis at the Catholic University of Rio de Janeiro, at the Federal University of Rio de Janeiro and at the Brazilian Institute of Geography and Statistics. This use of machines to process large amounts of data has generated a noticeable legacy in the geography practiced in the country today. Many study groups at universities in Brazil are currently focused on data processing with geographic information systems, and these techniques are also considerably widespread in government agencies.

Olivier Orain

I have no clues about digital geographies and big data because I have never worked on them. I try to understand geographies of the past, mainly in France, but increasingly in the USA, the UK, Russia, and Germany (because I can read the native languages). Working on living scholars is sometimes unsettling, as they try to control what you write about them. In addition, if they do not try to influence your work, there is still a chance that they will dislike what you say about them. Thus, I prefer the company of the dead, even though I have never been reluctant to write about contemporary topics – especially because they are more interesting for presentists, of which there are many throughout the scientific communities.

The most interesting topic for me would be to obtain a better understanding of the relationships among the different new approaches that have developed since the 1950s and 1960s among English speakers and the 1970s in France. I do not consider myself a historian of spatial analysis/GTQ exclusively, but rather someone who tries to understand the complexity of a given discipline or field and the competitive, cooperative, or polemic aspects that contribute to the structuring and diversification of the said field.

Luke R. Bergmann

Building on what I offered earlier, perhaps reading *Theoretical Geography* and engaging its milieu today offer what might be unexpected challenges and inspirations for new quantitative methodologies. Sciences find themselves at the center of politics and crises today; Bunge was a scientist whose work was avowedly political. GIS of recent decades has generally taken absolute space and location in an external coordinate system as central to its ontologies and epistemologies; *Theoretical Geography*, while cited by some in GIS as foundational, itself has sensibilities of space that may be equally close to many of those of contemporary human geography, with spaces that are multiple, emergent, and produced. The spaces of digital geographies are difficult to conceptualize, often resulting in retreat from landscapes to the use of graph and network representations alone; *Theoretical Geography* gestures toward spaces in between the network and the plane. Critical perspectives on the study of big data and data science have, inter alia, questioned whether empiricism has been overemphasized; *Theoretical Geography*, along with Waldo Tobler and others, offered paths for readers to reframe mapping not solely as an empirical inventory and visualization, but as sites for conceptual inquiry. *Theoretical Geography* is of its time and place, but its popular disciplinary legacy can sometimes be difficult to align with a reading of the text from the concerns of contemporary critical geography. As such, the relevance of *Theoretical Geography* and its milieu is potentially significant to contemporary challenges, perhaps unexpectedly so, given some of the received accounts of their legacies. It is inspirational to see how the concerns of the many present(s) in this current volume likewise inspire other re-readings of quantitative revolutions.

Matthew Hannah

Like the previous question, I interpret this one as concerning the legacy of the way we understand the development of quantitative methods. It seems to me that the traditional view of quantitative methods has been a barrier to more informed engagement with the new challenges posed by digitalization and big data. In this sense, a more contextually sensitive, nonabstract account of quantitative methods should encourage geographers to integrate statistical procedures into critical research programs. As the lives of many – not just in the Global North – come to be shaped in ever-more-central ways by big data and the digital, critical analysis needs to be based on some level of technical understanding and proficiency. Thus, again, quantitative methods – this time not just as purportedly free-floating tools but critically grasped in their entwinement with politics – need to be restored to a more prominent place in geography curricula.

Michiel van Meeteren

This relevance is a key point (Derudder and Van Meeteren 2019). Those who do not learn from history are doomed to repeat it. Geography wholeheartedly embraced quantitative methods after the quantitative revolution in the 1960s and got extensive experience in finding the limits of the understanding these methods can bring. We can learn from that encounter (Uitermark and Van Meeteren 2021). Equally important, other disciplines that now discover "urban science" can also learn from our history. At the same time, contemporary computational and data developments mean that we can do all the analyses they could only dream of in the 1960s. There is a fascinating amount of work to do in replicating research and amending theories from the quantitative era.

References

Bartels, D. (1980): Die konservative Umarmung der "Revolution". Zu Eugen Wirths Versuch in "Theoretischer Geographie". In *Geographische Zeitschrift*, 68(2), 121–131.

Brotton, J. (2012): *A History of the World in Twelve Maps*. London: Allen Lane.

Ceron, A. O.; Diniz, J. A. F. (1971): Tipologia da Agricultura: Questões Metodológicas e Problemas de Aplicação no Estado de São Paulo. In *Revista Brasileira de Geografia*, 32(3), 41–71.

Christaller, W. (1933): *Die zentralen Orte in Süddeutschland*. Jena: Gustav Fischer Verlag.

Christofoletti, A. (1982): As Perspectivas dos Estudos Geográficos. In Christofoletti, A. (Ed.): *Perspectivas da Geografia*. São Paulo: DIFEL, 11–36.

Derudder, B.; Van Meeteren, M. (2019): Engaging with "Urban Science." In *Urban Geography*, 40(4), 555–564.

Enyedi, G. (1984): Előszó. In Sikos, T. T. (Ed.): *Matematikai és Statisztikai Módszerek Alkalmazási Lehetőségei a Területi Kutatásokban*. Budapest: Akadémiai Kiadó, 9–10.

Faissol, S. (1960): XX Assembléia Geral do Conselho Nacional de Geografia. In *Boletim Geográfico, Rio de Janeiro*, 18(155), 187–189.

Feyerabend, P. (1975): *Against Method: Outline of an Anarchistic Theory of Knowledge*. London: New Left Books.

Gambi, L. (1971): Prefazione. In Mainardi, R. (Ed.): *Le Grandi Città Italiane*. Milano: Franco Angeli, 3–4.

Harvey, D. (1972): Revolutionary and Counter Revolutionary Theory in Geography and the Problem of Ghetto Formation. In *Antipode*, 4(2), 1–13.

Johnston, R.; Harris, R.; Jones, K.; Manley, D.; Wang, W. W.; Wolf, L. (2019): Quantitative Methods I: The World We Have Lost: Or Where We Started from. In *Progress in Human Geography*, 43(6), S. 1133–1142.

Kitchin, R.; Gleeson, J.; Dodge, M. (2013): Unfolding Mapping Practices: A New Epistemology for Cartography. In *Transactions of the Institute of British Geographers*, 38(3), S. 480–496.

Klaassen, L. H.; Van Dongen Torman, D. H.; Koyck, L. M. (1949): *Hoofdlijnen van de Sociaal-Economische Ontwikkeling der Gemeente Amersfoort van 1900–1970*. Leiden: Stenfert-Kroese.

Saey, P. (1968): A New Orientation in Geography. In *Bulletin De La Société Belge d'Etudes Géographiques/Tijdschrift Van De Belgische Vereniging Voor Aardrijkskundige Studies*, 37(1), 123–190.

Santos, M. (1978): *Por uma Geografia Nova*. São Paulo: Hucitec/Edusp.

Steigenga, W. (1950): Book Review of Klaassen et al. 1949, Hoofdlijnen van de Sociaal-Economische Ontwikkeling der Gemeente Amersfoort van 1900–1970. In *Tijdschrift voor Economische en Sociale Geografie*, 41(6), 153–154.

Uitermark, J.; Van Meeteren, M. (2021): Geographical Network Analysis. In *Tijdschrift voor Economische en Sociale Geografie*, 112(4), 337–350.

Ullman, E. L. (1954): Geography as Spatial Interaction. Reprinted in 1980 in Ullman, E. L.; Boyce, R. R. (Eds.): *Geography as Spatial Interaction*. Seattle, WA: University of Washington Press, 13–27.

Van Meeteren, M. (2016): *From Polycentricity to a Renovated Urban Systems Theory: Explaining Belgian Settlement Geographies*. PhD Thesis. Ghent: Ghent University.

Van Meeteren, M.; Sidaway, J. D. (2020): History of Geography. In *International Encyclopedia of Human Geography*. Amsterdam: Elsevier, volume 7, 37–44.

Virilio, P. (1984): *L'espace Critique*. Paris: Christian Bourgois.

Wirth, E. (1979): *Theoretische Geographie. Grundzüge einer theoretischen Kulturgeographie*. Stuttgart: Teubner (Teubner Studienbücher Geographie).

Index

Note: Page numbers in *italics* indicate a figure and page numbers in **bold** indicate a table on the corresponding page. Page numbers followed by "n" indicate a note.

Printed in the USA
CPSIA information can be obtained
at www.ICGtesting.com
LVHW010245190324
774517LV00064B/1719